SpringerBriefs in Applied Sciences and Technology

Thermal Engineering and Applied Science

Series Editor

Francis A. Kulacki, Department of Mechanical Engineering, University of Minnesota, Minneapolis, MN, USA

More information about this series at http://www.springer.com/series/10305

Sujoy Kumar Saha • Hrishiraj Ranjan
Madhu Sruthi Emani • Anand Kumar Bharti

Performance Evaluation Criteria in Heat Transfer Enhancement

 Springer

Sujoy Kumar Saha
Mechanical Engineering Department
Indian Institute of Engineering
Science and Technology, Shibpur
Howrah, West Bengal, India

Hrishiraj Ranjan
Mechanical Engineering Department
Indian Institute of Engineering
Science and Technology, Shibpur
Howrah, West Bengal, India

Madhu Sruthi Emani
Mechanical Engineering Department
Indian Institute of Engineering
Science and Technology, Shibpur
Howrah, West Bengal, India

Anand Kumar Bharti
Mechanical Engineering Department
Indian Institute of Engineering
Science and Technology, Shibpur
Howrah, West Bengal, India

ISSN 2191-530X   ISSN 2191-5318   (electronic)
SpringerBriefs in Applied Sciences and Technology
ISSN 2193-2530   ISSN 2193-2549   (electronic)
SpringerBriefs in Thermal Engineering and Applied Science
ISBN 978-3-030-20760-1   ISBN 978-3-030-20758-8   (eBook)
https://doi.org/10.1007/978-3-030-20758-8

This Springer imprint is published by the registered company Springer Nature Switzerland AG
The registered company address is: Gewerbestrasse 11, 6330 Cham, Switzerland

Contents

Nomenclature

A	Heat transfer surface area, m^2 or ft^2
A_c	Flow area at minimum cross section, m^2 or ft^2
A_{fr}	Frontal area, m^2 or ft^2
Am	Amplitude of temperature oscillation, °C
A_m	Mean conduction surface area, m^2 or ft^2
b	Distance between parting sheets, m or ft
B	External-to-internal surface area ratio, dimensionless
Ca	Capillary number
Co	Confinement number
cp	Specific heat capacity at constant pressure (J/kg K)
C_p	Specific heat, J/kg-K or Btu/lbm-°F
CPI	Coil per inch
d	Diameter
D_h	Hydraulic diameter, m or ft
d_i	Fin root diameter
d_i	Tube inside diameter, m or ft
d_o	Outer diameter
E	Electric field strength
E_1	Parameter
E_2	Parameter
f	Cycle frequency of temperature oscillation, Hz
fA	Product of f and Am, °C/s
FG	Fixed geometry
FN	Fixed cross-sectional flow area
g	Gravitational acceleration, m/s^2 or ft/s^2
G	Mass velocity, $kg/s\text{-}m^2$ or $lbm/s\text{-}ft^2$
G_{fr}	Mass velocity based on heat exchanger frontal area, $kg/s\text{-}m^2$ or $lbm/s\text{-}ft^2$
h	Heat transfer coefficient, $W/m^2\text{-}K$ or $Btu/hr\text{-}ft^2\text{-}°F$
j	Heat transfer factor, $StPr^{2/3}$, dimensionless
j	Colburn factor

k	Thermal conductivity, W/m-K or Btu/hr-ft^2-°F
L	Length of flow path in heat exchanger, m or ft
LMTD	Log-mean temperature difference, K or °F
n	Length to radius ratio of duct
N	Number of tubes in each pass of heat exchanger, dimensionless
NTU	Number of heat transfer units, UA/C_p W_{min}, dimensionless
Nu	Nusselt number
P	Pumping power, W or HP
P	Roughness axial pitch, m or ft
ΔP	Pressure drop, Pa or lbf/ft^2
Pr	Prandtl number
q	Heat flux, W/m^2 or Btu/h-ft^2
Q	Heat transfer rate, W or Btu/h
Q_{av}	Available heat transfer rate, W or Btu/h
r	Resistance ratios
R_0	1/$h_o A_o$, m^2-K/W or ft^2 h-°F/Btu
r_c	Radius of curvature for square duct with rounded corners (m)
Re	Reynolds number: smooth or rough tube (du/v); internally finned tube ($D_h u/v$), dimensionless
R_f	Fouling factor, m^2-K/W or ft^2-h-°F/Btu
R_i	1/$h_i A_i$, m^2-K/W or ft^2-h-°F/Btu
St	Stanton number, $h/C_P G$ dimensionless
T	Temperature, K or °F
ΔT	Average temperature difference between hot and cold streams, K or °F
ΔT_i	Temperature difference between inlet hot and cold streams, K or °F
t	Tube wall thickness, m or ft
u	Average flow velocity, m/s or ft/s
U	Overall heat transfer coefficient, W/m^2-K or °F
V	Heat exchanger volume, $A_{ft} L$, m^3 or ft^3
VG	Variable geometry
W	Flow rate, kg/s or lbm/s
x_i	Inlet vapor quality to heat exchanger, dimensionless
Δx	Vapor quality change, dimensionless
τ	Nondimensional temperature difference, ϕ/T_o

Subscripts

a	Air
ave.	Average
c	Cavity
c	Per refrigerant circuit
CHF	At critical heat flux condition
eq	Equivalent

exit	At the exit section
f	Evaluated at film temperature
G	Gaseous phase
I	Inertia
inlet	At the inlet section
M	Evaporation momentum
max	Maximum
min	Minimum
o	Outside of the tube
ONB	Onset of nucleate boiling
p or s	Plain or smooth surface geometry
r	Receding
S	Surface tension
Sat	Saturation
Sub	Subcooling
τ	Shear (viscous)
V	Vapor
w	Wall condition
wall	Channel wall
$Y = \infty$	For an untwisted or straight tube

Greek Symbol

β	Volume flow fraction
δ_t	Thermal boundary layer thickness, m
λ	Parameter in empirical correlations
μ	Dynamic viscosity
ρ	Density
ρ_m	Average density
σ	Area ratio A_{fr}/A_c, surface tension
ν	Kinematic viscosity, m^2/s or ft^2/s
ν_g	Saturated vapor specific volume, m^3/kg or ft^3/lbm
ν_l	Specific volume of liquid, m^3/kg or lbm/ft^3

Chapter 1
Introduction

1.1 Performance Evaluation Criteria for Heat Exchangers

A quantitative assessment of performance in general and that of a heat exchanger in particular may be obtained by comparing the performance of an enhanced surface with that of the corresponding plain, i.e. smooth surface [1–6].

The performance of the enhanced surface is evaluated on the basis of the following:

1. Size reduction: With constant heat exchange, the length of the heat exchanger may be reduced and a smaller heat exchanger is obtained.
2. Increased UA gives reduced ΔT_m. For constant tube length and heat duty, ΔT_m is reduced. This gives increased thermodynamic process efficiency and savings in operating cost. Increased UA/L results in increased heat exchange rate for fixed fluid inlet temperature and L.
3. Enhanced heat exchanger operates at a smaller velocity than that in smooth (plain) surface and the enhanced surface causes reduced pumping power for fixed heat duty. However, this requires increased frontal area. Operating conditions like fluid flow rate and entering fluid temperature are also important.

The basic performance of an enhanced surface for single-phase heat transfer is evaluated by the j factor ($j = StPr^{2/3}$) and friction factor versus Reynolds number plot. The pumping power constraint is a very important consideration for evaluating the performance of an enhanced surface in single-phase flow.

Several performance evaluation criteria have been found in the literature. The effects of both the fluid streams are important for the performance evaluation. The PEC analysis for enhancement of a two-phase flow (boiling and condensation) is different from that of a single-phase flow since, in case of two-phase flow, the local saturation temperature of the fluid is reduced by the pressure drop of the two-phase fluid and the driving potential for heat transfer also gets affected. The heat exchanger

S. K. Saha et al., *Performance Evaluation Criteria in Heat Transfer Enhancement*,
SpringerBriefs in Applied Sciences and Technology,
https://doi.org/10.1007/978-3-030-20758-8_1

design objects may be to have smaller heat exchanger or to have improved thermo-dynamic process efficiency. Size reduction of a heat exchanger is often a very important criterion, since this results in cost reduction and use of smaller volume of expensive refrigerant in the refrigeration industry. Also, other PEC may be useful from the life cycle costing point of view; for example reduced refrigeration con-densers and evaporators result in reduced compressor power costs. Performance evaluation criteria including cost considerations have been studied by Hewitt et al. [7], Evans and Churchill [8], Bergles et al. [5, 6], Webb [9], Brown [10], El-Sayed [11] and Fraas [12].

Upgrading the capacity of an existing heat exchanger may also be a preferred objective. A practical enhanced surface is required to provide the desired heat transfer enhancement and meet the required flow rate and pressure drop constraints. A surface geometry, which gives a given heat duty with the lowest pressure drop, is definitely preferred.

Our objective is to determine how the enhanced surface affects the performance of the heat exchanger for a given set of operating conditions and design constraints. The four performance objectives, for fixed heat exchanger flow rate and entering fluid temperature, are:

(a) Reduced heat transfer surface area for fixed heat duty and pressure drop
(b) Reduced LMTD for fixed heat duty and surface area
(c) Increased heat duty for fixed surface area
(d) Reduced pumping power for fixed duty and surface area

Several different objects lead to different results like smaller heat exchanger size and reduced capital cost; reduced operating cost; improved system thermodynamic efficiency, yielding lower system operating cost; and increased heat exchange capacity of a given heat exchanger. More expensive heat exchanger with heat transfer surface will have high operating cost savings, even though the enhanced heat exchanger may have same total tubing length.

One performance evaluation criterion is determined considering one of the operating parameters for the performance objective keeping the remaining design constraints unchanged. Major operating variables are heat transfer rate, fluid pumping power, heat exchanger flow rate and fluid velocity or flow frontal area.

Fan et al. [13] presented a plot for performance evaluation of various heat transfer enhancement techniques which are aimed at energy saving as shown in Fig. 1.1. The plot co-ordinates were log-log based. The two co-ordinates of the plot are represented by $\frac{Nu_e}{Nu_o}$ and $\frac{f_e}{f_o}$ where "e" represents enhanced tube and "o" is for the smooth tube. All the lines for different enhancement techniques were passing through the point (1,1). The plot has been divided into four regions for better understanding of the enhancement criteria. The increase in both $\frac{Nu_e}{Nu_o}$ and $\frac{f_e}{f_o}$ can be observed in region 1. But the increase in $\frac{Nu_e}{Nu_o}$ is quite less compared to the increase in $\frac{f_e}{f_o}$. This implies that the enhancement technique is ineffective.

The increase in heat transfer rate for enhanced tube over that in a smooth tube at constant pumping power consumption can be observed in region 2. Region

Fig. 1.1 A performance evaluation plot for enhancement technique oriented for energy saving [13]

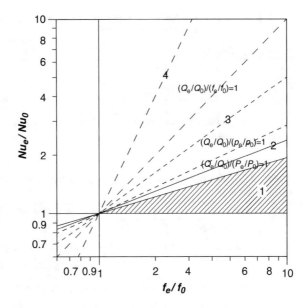

Fig. 1.2 Sketch of repeated-rib roughness geometry [1]

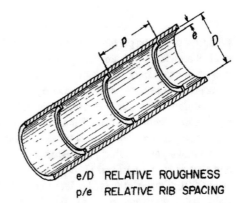

e/D   RELATIVE ROUGHNESS
p/e   RELATIVE RIB SPACING

3 represents relative increase in heat transfer rate in augmented tube for identical pressure drop. The significant increase in heat transfer rate as compared to that in friction factor, at the same flow rate, is observed in region 4. Thus, for any heat transfer augmentation technique, region 4 is desirable.

The evaluation of rough surfaces for heat transfer enhancement in heat exchangers has been done by Webb and Eckert [1]. The tube with repeated rib roughness has been shown in Fig. 1.2. The performance of rough tubes relative to the smooth tube at constant wall temperature boundary condition has been shown in Table 1.1. Also, the performance of rough tubes for heat transfer between two fluids across the wall has been shown in Table 1.2.

Possible PEC relations are obtained from the design constraints on flow rate and flow velocity. Increased friction factor requires reduced flow velocity for fixed

Table 1.1 Relative performance for rough tubes with specified wall temperature [1]

Case	Description	Constraints	Parameter of interest	Relative mass velocity (G^*)
A	Reduced surface area	$\frac{P}{P_s} = \frac{Q}{Q_s} = 1$	$\frac{A}{A_s} = \frac{(f/f_s)^{1/2}}{(St/St_s)^{3/2}}$	$G^* = \frac{(f/f_s)^{\frac{1}{2}}}{(St/St_s)^{\frac{1}{2}}}$
B	Increased heat transfer	$\frac{P}{P_s} = \frac{A}{A_s} = 1$	$\frac{K}{K_s} = \frac{St/St_s}{(f/f_s)^{1/3}}$	$G^* = (f/f_s)^{\frac{1}{3}}$
C	Reduced friction power	$\frac{Q}{Q_s} = \frac{A}{A_s} = 1$	$\frac{P}{P_s} = \frac{f/f_s}{(St/St_s)^3}$	$G^* = St/St_s$

Table 1.2 Relative performance for rough tubes when heat is exchanged between two fluids across a pipe wall [1]

Case	Description	Constraints	Parameter of interest	Relative mass velocity (G^*)
A	Reduced surface area	$\frac{P}{P_s} = \frac{Q}{Q_s} = 1$	$\frac{A}{A_s} = \frac{G^*(St_s/St)+r}{1+r}$	$\frac{(f/St)}{(f_s/St_s)} = (G^*)^2(1+r) - \frac{r(f/f_s)}{G^*}$
B	Increased heat transfer	$\frac{P}{P_s} = \frac{A}{A_s} = 1$	$\frac{K}{K_s} = \frac{1+r}{(f/f_s)^{1/3}(St/St_s)+r}$	$G^* = \left(\frac{f}{f_s}\right)^{\frac{1}{3}}$
C	Reduced friction power	$\frac{Q}{Q_s} = \frac{A}{A_s} = 1$	$\frac{P}{P_s} = (f/f_s)/(St/St_s)^3$	$G^* = \frac{St}{St_s}$

pumping power or pressure drop constraint. Fixed flow rate requires increased flow control area for fixed pumping power design constraints. On the other hand, reduced mass flow rate may also keep constant flow frontal area at reduced velocity. For reduced flow rate, the system must operate at higher thermal effectiveness to keep the required heat duty. Performance potential of the enhanced surface may be significantly reduced, if the design thermal effectiveness is sufficiently high. When the heat exchanger flow rate is specified, the flow rate reduction is not permitted.

Andrews and Fletcher [14], Brown [10], Bergwerk [15], El-Sayed [11], Heggs and Stones [16], Sano and Usui [17], Norris [18], Sparrow and Liu [19], Zimparov et al. [20], Yılmaz et al. [21], Jiang et al. [22], Zukauskas [23], Zhang et al. [24], Zaherzadeh and Jagadish [25], Yapici [26], Wood et al. [27], William et al. [28], Webb et al. [29], Webb [30] and Sorlie [31] have studied the performance evaluation criteria for various heat transfer augmentation techniques.

References

1. Webb RL, Eckert ER (1972) Application of rough surfaces to heat exchanger design. Int J Heat Mass Transfer 15(9):1647–1658
2. Webb RL (1981) Performance evaluation criteria for use of enhanced heat transfer surfaces in heat exchanger design. Int J Heat Mass Transfer 24(4):715–726
3. Bergles AE (1981) Applications of heat transfer augmentation. In: Kakac S, Bergles AE, Mayinger F (eds) Heat exchangers: thermal hydraulic fundamentals and design. Hemisphere, Washington, DC

4. Webb RL, Bergles AE (1983) Performance evaluation criteria for selection of heat transfer surface geometries used in low Reynolds number heat exchangers. In: Kakac S, Shah RK, Bergles AE (eds) Low Reynolds number flow heat exchangers. Hemisphere, Washington, DC
5. Bergles AE, Blumenkrantz AR, Taborek, J (1974a) Performance evaluation criteria for enhanced heat transfer surfaces. In: Proc. 4th int. heat transfer conf., vol 2, pp 239–243
6. Bergles AE, Bunn RL, Junkhan GH (1974b) Extended performance evaluation criteria for enhanced heat transfer surfaces. Lett Heat Mass Transfer 1:113–120
7. Hewitt GF, Shires GL, Bott RL (1994) Process heat transfer. Begell House, New York, pp 11–42
8. Evans LB, Churchill SW (1962) Chem Eng Prog Symp Ser 58(10):55
9. Webb RL (1982) Performance, cost effectiveness, and water-side fouling considerations of enhanced tube heat exchangers for boiling service with tube-side water flow. Heat Transfer Eng 3(3–4):84–98
10. Brown TR (1986) Use these guidelines for quick preliminary selection of heat-exchanger type. Chem Eng 93:107–108
11. El-Sayed M (1999) Revealing the cost-efficiency trends of the design concepts of energy-intensive systems. Energy Convers Manag 40:1599–1615
12. Fraas AP (1989) Heat exchanger design. John Wiley, New York, pp 216–228
13. Fan JF, Ding WK, Zhang JF, He YL, Tao WQ (2009) A performance evaluation plot of enhanced heat transfer techniques oriented for energy-saving. Int J Heat Mass Transfer 52 (1–2):33–44
14. Andrews MJ, Fletcher LS (1996) Comparison of several heat transfer enhancement technologies for gas heat exchangers. J Heat Transfer 118:897–902
15. Bergwerk W (1963) The utilisation of research data in heat exchanger design. Proc IMechE 178:55–81
16. Heggs PJ, Stones PR (1980) Improved design methods for finned tube heat exchangers. Trans IChemE 58:147–154
17. Sano Y, Usui H (1982) Evaluation of heat transfer promoters by the fluid dissipation energy. Heat Transfer Jpn Res 11:91–96
18. Norris RH (1970) Some simple approximate heat transfer correlations for turbulent flow in ducts with rough surfaces, in augmentation of convective heat and mass transfer. ASME 16:16–26
19. Sparrow EM, Liu CH (1979) Heat-transfer, pressure-drop and performance relationships for in-line, staggered, and continuous plate heat exchangers. Int J Heat Mass Transfer 22:1613–1625
20. Zimparov VD, Vulchanov NL, Delov LB (1991) Heat transfer and friction characteristics of spirally corrugated tubes for power plant condensers—1. Experimental investigation and performance evaluation. Int J Heat Mass Transfer 34(9):2187–2197
21. Yılmaz M, Yapıcı S, Çomaklı Ö, Şara ON (2002) Energy correlation of heat transfer and enhancement efficiency in decaying swirl flow. Heat Mass Transfer 38(4–5):351–358
22. Jiang PX, Fan MH, Si GS, Ren ZP (2001) Thermal-hydraulic performance of small cale microchannel and porous-media heat exchangers. Int J Heat Mass Transfer 44:1039–1051
23. Zukauskas A (1986) Heat transfer augmentation in single-phase flow. In: Proc. 8th int. heat transf. conf., vol 1, pp 47–57
24. Zhang YM, Gu WZ, Han JC (1994) Heat transfer and friction in rectangular channels with ribbed or ribbed-grooved walls. J Heat Transfer 116:58–65
25. Zaherzadeh NH, Jagadish BS (1975) Heat transfer in decaying swirl flows. Int J Heat Mass Transfer 18:941–944
26. Yapici S (1999) Energetic correlation of local mass transfer in swirling pipe flow. Ind Eng Chem Res 38:1712–1717
27. Wood AS, Tupholme GE, Bhatti MIH, Heggs PJ (1996) Performance indicators for steady-state heat transfer through fin assemblies. J Heat Transfer 118:310–316

28. William F, Pirie MAM, Warburton C (1970) Heat transfer from surfaces roughened by ribs. In: Bergles AE, Webb RL (eds) Augmentation of convective heat and mass transfer. ASME, New York, pp 36–43
29. Webb RL, Eckert ERG, Goldstein RJ (1971) Heat transfer and friction in tubes with repeated-rib roughness. Int J Heat Mass Transfer 14:601–617
30. Webb RL (1994) Principles of enhanced heat transfer. John Wiley, New York
31. Sorlie T (1962) Three-fluid heat exchanger design theory, counter and parallel flow. Tr. No. 54. Department of Mechanical Engineering, Stanford University, CA

Chapter 2
Single-Phase Flow Performance Evaluation Criteria

The performance evaluation criteria classified into different groups have been shown in Fig. 2.1. Table 2.1 shows some PEC for single-phase fluid flow inside tubes.

2.1 Fixed Geometry (FG) Criteria

2.1.1 FG-1

The cross-sectional flow area and tube length remain constant. Basic geometry, like tube envelope diameter, tube length and number of tubes for in-tube flow, remains same in case of enhanced geometry as that in case of the smooth surfaces.

2.1.2 FG-2

The enhanced surface would operate at the same pumping power but the enhanced exchanger would normally operate at reduced flow rate.

2.1.3 FG-3

In this case the enhanced geometry seeks reduced pumping power for fixed heat duty.

© The Author(s), under exclusive license to Springer Nature Switzerland AG 2020
S. K. Saha et al., *Performance Evaluation Criteria in Heat Transfer Enhancement*,
SpringerBriefs in Applied Sciences and Technology,
https://doi.org/10.1007/978-3-030-20758-8_2

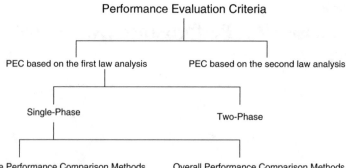

Fig. 2.1 General classification of performance evaluation criteria (Yilmaz, 2005)

Table 2.1 Performance evaluation criteria for single-phase heat exchanger system

		Fixed				
Case	Geometry	W	P	Q	ΔT_i	Objective
FG-1a	N,L^a	×			×	$\uparrow Q$
FG-1b	N,L	×		×		$\downarrow \Delta T_i$
FG-2a	N,L		×		×	$\uparrow Q$
FG-2b	N,L		×	×		$\downarrow \Delta T_i$
FG-3	N,L			×	×	$\downarrow P$
FN-1	N	×	×	×	×	$\downarrow L$
FN-2	N	×		×	×	$\downarrow L$
FN-3	N	×		×	×	$\downarrow P$
VG-1		×	×	×	×	$\downarrow NL$
VG-2a	N,L^b	×	×		×	$\uparrow Q$
VG-2b	N,L^b	×	×	×		$\downarrow \Delta T_i$
VG-3	N,L^b	×	×	×	×	$\downarrow P$

[a]N and L are constant in all FG cases
[b]The product NL is constant for VG-2 and VG-3

2.2   Fixed Cross-Sectional Flow Area (FN)

The length of the heat exchanger is a variable. The enhanced geometry has reduced surface area (FN-1) or reduced pumping power (FN-2) for constant heat duty.

2.3   Variable Geometry (VG) Criteria

A heat exchanger may have required thermal duty with a specified flow rate. Tube-side velocity is reduced and the enhanced surface has a higher friction characteristics. Flow area is increased to keep the flow rate constant. The situation needs a greater number of parallel flow circuits. Since the exchanger flow rate remains constant, operation at higher thermal effectiveness is avoided.

Calculation of performance requires algebraic relations to quantify the objective function and constraints. PEC analysis varies for the case of a prescribed wall temperature or a prescribed wall flux. Algebraic relation relative to a smooth surface may be developed for the same fluid operating temperature. The fluid properties need not be calculated from the equations.

The PEC may be formulated as the following:

$$
\frac{\left(\dfrac{hA}{h_s A_s}\right)}{\left(\dfrac{P}{P_s}\right)^{1/3}\left(\dfrac{A}{A_s}\right)^{2/3}} = \frac{j/j_s}{\left(f/f_s\right)^{1/3}} \tag{2.1}
$$

One of the variables, $\frac{hA}{h_s A_s}$, $\frac{P}{P_s}$ and $\frac{A}{A_s}$, may be set as an objective function and the remaining two may be set as operating conditions, treating the parameter without any subscript same as that with subscript "s".

The correlations for smooth surfaces and those for enhanced surfaces must be known. The heat exchanger flow rate, the flow frontal area and the heat transfer coefficient on the outer and inner tube surfaces for the smooth surface heat exchanger must be known.

A simple surface performance comparison may be calculated as follows:
Assumption:

(a) A shell-and-tube heat exchanger is considered.
(b) Enhancement is applied to the tube side.
(c) Total thermal resistance is on the tube side.
(d) Tube-for-tube replacement occurs, i.e. there is fixed flow frontal area.

For constant flow rate:
Following may be used to evaluate the performance of the enhanced surface:

$$\left(\frac{A}{A_s}\right) = \left(\frac{f}{f_s}\right)^{\frac{1}{2}} \left(\frac{j_s}{j}\right)^{\frac{3}{2}}$$ (2.2)

$$\left(\frac{G}{G_s}\right) = \left(\frac{j}{j_s} \cdot \frac{f_s}{f}\right)^{\frac{1}{2}}$$ (2.3)

$$\left(\frac{W}{W_s}\right) = 1 = \left(\frac{G}{G_s}\right)\left(\frac{N}{N_s}\right)$$ (2.4)

$$\left(\frac{A}{A_s}\right) = \left(\frac{N}{N_s}\right) \cdot \left(\frac{L}{L_s}\right)$$ (2.5)

For fixed number of tubes in the exchanger, the flow frontal area remains constant. The reduced velocity results in reduced flow rate. This heat exchanger must operate at higher thermal effectiveness, and greater surface area is required. For reduced heat exchange flow rate, there is less benefit of enhancement. The effect of reduced flow rate on the performance improvement must be accounted for. For design purpose, the flow rate reduction must be avoided. The different PEC as defined by Bergles et al. [1] has been presented in Table 2.2 along with their constraints, objectives and benefits.

Garcia et al. [2] studied the heat transfer enhancement performance of wire-coil inserts. They considered the laminar, transition and turbulent flow regimes. The performance has been evaluated by the R_3 criteria and is presented in Fig. 2.2. They reported that at higher Prandtl number the enhancement rate was high in laminar and transition regimes. However, the effect of Prandtl number was almost negligible in turbulent flow. According to the R_3 evaluation criterion, the wire coils were ineffective in laminar flow region. But the transition from laminar to turbulent started early. The performance of the wire coils was found to be quite effective in the transition region with maximum R_3 being 200%.

Vicente et al. [3] studied the heat transfer and pressure drop characteristics of flow through tube having helical dimples. The flow was considered to be low Reynolds turbulent flow. They experimentally studied the performance of ten different tubes. The geometrical details of the tubes have been shown in Table 2.3. They used the R_1, R_3 and R_5 performance evaluation criteria as defined by Bergles et al. [1]. The performance of the helically dimpled tubes using R_3 and R_5 PEC has been shown in Figs. 2.3 and 2.4, respectively. They concluded that the dimples with greater depth showed superior performance. They observed an increase in

Table 2.2 Summary of performance criteria evaluations [1]

Variable / Criterion number	Fixed					Objective				Performance ratio equations	Specifications
	Basic geometry	Flow rate	Pressure drop	Pumping power	Heat duty	Heat transfer increase	Pumping power reduction	Exchanger size reduction	Total cost		
1	X	X				X				$R_1 = \left(\frac{h_a}{h_o}\right)_{D,L,N,\dot{m},T_m',\Delta T} = \frac{q_a}{q_o}$	(1) It is applied in heat exchanger in which there are geometrical constraints, and the stream flow rate is fixed by the general process conditions (2) The geometrical constraint involves keeping the exchanger length, tube nominal diameter and number of tubes constant
2	X		X			X				$R_2 = \left(\frac{h_a}{h_o}\right)_{D,L,N,\Delta p,T_m',\Delta T} = \frac{q_a}{q_o}$	(1) It is of interest when operation occurs on the flat portion of the pump head-flow curve
3	X			X		X				$R_3 = \left(\frac{h_a}{h_o}\right)_{D,L,N,P,T_m',\Delta T} = \frac{q_a}{q_o}$	(1) It is relevant to the pumping power or operating cost, and is utilised when the cost of fluid pumping is a major consideration (2) It is called as *enhancement efficiency* or heat transfer efficiency
4	X				X		X			$R_4 = \left(\frac{P_a}{P_o}\right)_{D,L,N,q,T_m',\Delta T}$	(1) It is utilised if the aim is to reduce the pumping power while keeping the heat transfer rate same

(continued)

Table 2.2 (continued)

Variable	Fixed					Objective				Performance ratio equations	Specifications
	Basic geometry	Flow rate	Pressure drop	Pumping power	Heat duty	Heat transfer increase	Pumping power reduction	Exchanger size reduction	Total cost		
											(2) The augmented tube heat transfer data should be consistent with the smooth tube data (3) It will be difficult to keep the heat duty constant for this comparison since ΔT will usually be altered
5				X	X			X		$R_5 = \left(\dfrac{A_a}{A_o}\right)_{q,P,D_i,T_m,\Delta T} - \dfrac{h_o}{h_a}$	(1) The size reduction may be brought about via reducing the length or the number of tubes, with an increased wetted perimeter being possible with the augmented tube (2) Implementation of this criterion usually requires the pump to be changed
6			X		X			X		$R_6 = \left(\dfrac{A_a}{A_o}\right)_{q,\Delta p,D_i,T_m,\Delta T} - \dfrac{h_o}{h_a}$	(1) It is of interest when operation occurs on the flat portion of the pump head-flow curve
7		X			X			X		$R_7 = \left(\dfrac{A_a}{A_o}\right)_{q,m,D_i,T_m,\Delta T} - \dfrac{h_o}{h_a}$	(1) It is of interest when the heat duty and flow rate are fixed for the objective of the reducing exchanger size (2) There is a trade-off between length and number of tubes

No.									Criterion	Remarks
8			X		X		X	X		$R_8 = \left(\dfrac{A_a}{A_o}\right)_{q,m,\Delta T,D,T_m,\Delta T} - \dfrac{h_o}{h_a}$	(3) Implementation of this criterion generally requires a new pump to be used (1) It is used when the flow rate and heat duty are fixed, imposed by process conditions, and it is desired to use the same pump
9						X			X	$R_9 = \left(\dfrac{h_a}{h_o}\right)_{D,L,N,C,T_m,\Delta T} - \dfrac{q_a}{q_o}$	(1) It is used when cost consideration is the deciding factor (2) If the basic cost information is available, different methods may be used for the best geometry for the specified cost constraint

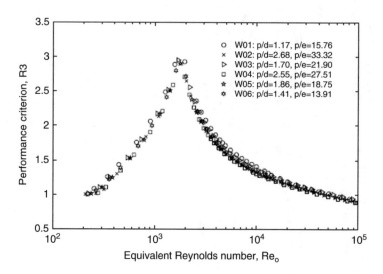

Fig. 2.2 Performance evaluation criterion R3 vs. equivalent smooth tube Reynolds number Re_0 for wire coils W01–W06 [2]

Table 2.3   Geometry of test tubes [3]

Tube no.	d (mm)	h (mm)	p (mm)	l (mm)	h/d	d^2/pl
01	16.0	1.33	13.0	8.85	0.0831	2.225
02	16.0	1.58	13.1	8.99	0.0988	2.175
03	16.0	1.91	13.8	8.89	0.1194	2.085
04	16.0	1.28	14.6	8.91	0.0800	1.975
05	16.0	1.83	14.5	9.02	0.1144	1.964
06	16.0	1.59	17.2	9.02	0.0994	1.652
07	16.0	1.84	16.6	9.06	0.1150	1.700
08	16.0	1.87	16.8	8.90	0.1169	1.709
09	16.0	1.55	10.9	8.90	0.0969	2.646
10	16.0	1.64	11.4	8.76	0.1025	2.575

heat transfer rate by 20–110% over that of smooth tube according to the R_3 criterion. Also, according to the R_5 criterion, the reduction in heat exchanger surface area of about 80–20% over that of plain tube has been reported.

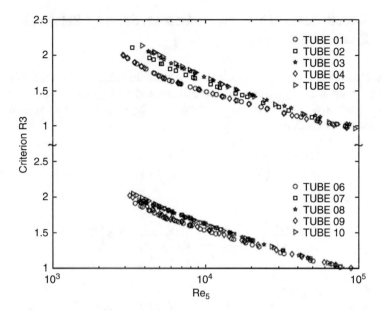

Fig. 2.3 R3 factor vs. *Re* [3]

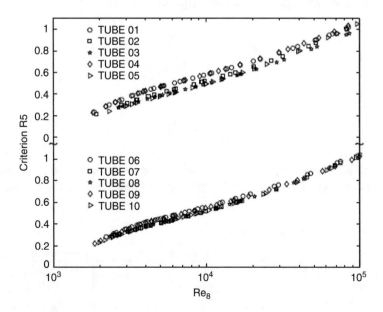

Fig. 2.4 R5 factor vs. *Re* [3]

2.4 Effectiveness, Thermal Resistance, St and f Relations, Reduced Flow Rate

The evaluation of performance, which accounts for the thermal resistance of both hot and cold fluid streams, wall thermal impedance and fouling resistance, is bit more difficult and involved. Following equations may be used for the analysis of such performance evaluation:

$$\left(\frac{P}{P_s}\right) = \left(\frac{f}{f_s}\right) \cdot \left(\frac{A}{A_s}\right) \cdot \left(\frac{G}{G_s}\right)^3 \tag{2.6}$$

$$\left(\frac{UA}{U_sA_s}\right) = \frac{1+\beta_s}{\left(\frac{St_s}{St}\right)\left[\left(\frac{f}{f_s}\right)\left(\frac{P}{P_s}\right)\left(\frac{A_s}{A}\right)^2\right]^{1/3}} + \beta\left(\frac{A_s}{A}\right) \tag{2.7}$$

where β and β_s are given in Table 2.4.

Equations (2.6) and (2.7) are equally applicable for tube-tube enhancement as well as shell-side enhancement or air-side enhancement of a fin-and-tube heat exchanger, with all terms in Eqs. (2.6) and (2.7) being well defined.

The tube diameter of the enhancement exchanger may be different from that of the reference smooth-tube exchanger. The definition of geometrical parameters for air-side extended surfaces is more involved. For the tube-side enhancement in a shell-and-tube exchanger, the following equations may be used:

$$\left(\frac{UA}{U_sA_s}\right) = \frac{1+\beta}{\left(\frac{St_s}{St}\right)\left(\frac{f}{f_s}\right)^{1/3}} + \beta \tag{2.8}$$

$$\left(\frac{G}{G_s}\right) = \left(\frac{f_s}{f}\right)^{1/3} \tag{2.9}$$

Figure 2.5 shows the performance benefits of tube-side transverse-rib roughness in a condenser in relation to a condenser having a smooth inside tube surface.

Table 2.4 Dimensionless ratios to be used in analysis

Definition	Reference exchanger	Enhanced exchanger
Outer surface conductance	$r_{cr} = h_sA_s/h_{os}A_{os}$	$r_o = h_sA_s/h_oA_o$
Metal resistance	$r_{ws} = h_stA_s/kA_m$	$r_w = htA/kA_m$
Fouling resistance	$r_{fs} = h_sR_f$	$r_f = hR_f$
Composite resistance	$\beta_s = r_{os} + r_{ws} + r_{fs}$	$\beta = r_o/E_{ho} + r_wA/A_s + r_f$

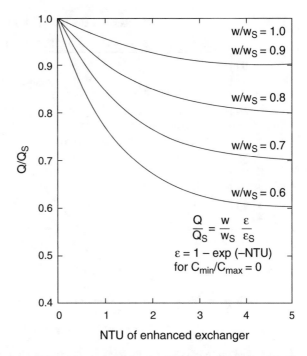

Fig. 2.5 Effect of reduced flow rate on Q/Q_s for $UA/U_sA_s = 1$ (Webb 1981a)

A given enhancement type needs several possible characteristic dimensions and the nominal tube diameter or hydraulic diameter is not sufficient for a generalised correlation in the form of St and f versus Re. So the correlations are not versatile. Bergles [4] shows the way of using single-tube data to a "scaled-up" situation that involves flow outside a large number of tubes. This is rather very important for two-phase heat transfer like boiling and condensation than for single-phase heat transfer.

2.5 Flow over Finned Tube Banks

The effectiveness of enhanced heat exchanger may be different from that of smooth exchangers.

NTU may be obtained from the following equations:

$$\left(\frac{Q}{Q_s}\right) = \left(\frac{W}{W_s}\right)\left(\frac{\varepsilon}{\varepsilon_s}\right)\left(\frac{\Delta T_i}{\Delta T_{is}}\right) \tag{2.10}$$

When inlet temperature is fixed, $\Delta T_i = \Delta T_{is}$ and

$$\left(\frac{Q}{Q_s}\right) = \left(\frac{W}{W_s}\right)\left(\frac{\varepsilon}{\varepsilon_s}\right) \tag{2.11}$$

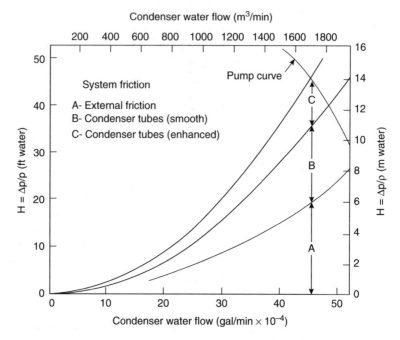

Fig. 2.6 Illustration of balance point between pump curve and system resistance for smooth and enhanced tubes in steam condenser of a nuclear plant (Webb and Kim 2005)

$$\text{NTU} = (\text{NTU})_s \left(\frac{UA}{U_s A_s} \right) \left(\frac{W_s}{W} \right) \qquad (2.12)$$

Additional surface area is required to compensate for the reduced LMTD since thermal effectiveness is higher for the operation of FN exchanger.

Figure 2.6 shows the $\left(\frac{Q}{Q_s} \right)$ versus NTU of enhanced exchanger curve for the shell-side condensation. There may be benefits with reduced flow rate. A smaller pump, reduced pressure drop in the supply pipe and header, or smaller supply pipe size is the example.

The PEC may be for flow normal to the bare or finned tube banks. The benefits of an enhanced surface may be lost if the heat exchanger is operated with $\left(\frac{W}{W_s} \right) < 1$. With reduced flow rate, the LMTD is reduced and additional can only compensate for the reduced LMTD.

For constant pumping power, the greatest performance benefit of an enhanced surface accrues with the use of VG criteria. Enhanced surfaces perform better in new designs, with fixed flow area, than in the case of the older designs. The flow frontal area increases over that of the corresponding smooth tube design.

Gholami et al. [5] worked for evaluating the thermal-hydraulic performance in tube bank compact heat exchanger with innovative design of corrugated fin-and-oval

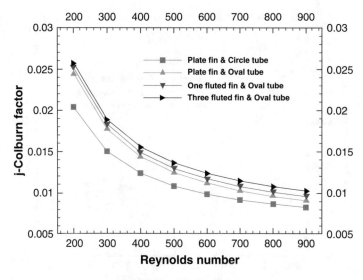

Fig. 2.7   Variation of *j*-Colburn factor vs. Reynolds number [5]

tubes. The evaluation of thermal-hydraulic performance criteria was carried out for
Reynolds number in the range of 200–900 by using computational fluid dynamic
method. The dimensionless quantity *j*-Colburn shows a relation between convective
heat transfer coefficient, fluid properties, flow condition and geometry. The outcome
of simulation graph between *j*-Colburn factor and Reynolds number is plotted in
Fig. 2.7. The simulated results showed that fin configuration has a profound effect on
the performance of hydrothermal characteristics. Figure 2.7 shows that *j* factor
decreases with increasing Reynolds number and increases with increasing number
of curvature regions for all the combination of fins and tubes. Figure 2.8 depicts
variation of overall performance (JF) with Reynolds number and the graph clearly
reveals that three corrugated fins have superior performance compared to conventional design of fin-and-tube compact heat exchangers (FTCHE). Table 2.5 shows
the governing equations used for numerical simulation in the FTCHE.

Overall thermal-hydraulic performance:

$$\mathrm{JF} = \frac{j}{\sqrt[3]{f}} \tag{2.13}$$

Extended performance evaluation criteria (ExPEC) was used by Petkov et al. [6]
to investigate the performance characteristics of single-phase fully developed laminar flow through a bundle of non-circular ducts having rectangular, isosceles
triangular, elliptical, trapezoidal and hexagonal cross section. Bundle of circular
tubes was considered as a reference heat transfer unit. The performance characteristics of these non-circular ducts have been compared with the reference circular
tube. Constant wall temperature was the thermal boundary condition. Figure 2.9
shows the variation of dimensionless heat transfer rate with the ratio of shape factors

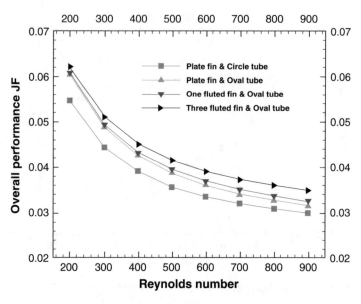

Fig. 2.8 Variation of JF factor vs. Reynolds number [5]

Table 2.5 Governing equation used for numerical simulation in the FTCHE [5]

Continuity equation: $\frac{\partial(\rho u_i)}{\partial x_i} = 0.0$

Momentum equation: $\frac{\partial}{\partial x_i}\left(\rho u_i u_j\right) = \frac{\partial}{\partial x_i}\left(\mu \frac{\partial u_j}{\partial x_i}\right) - \frac{\partial p}{\partial x_j}$

Energy equation: $\frac{\partial}{\partial x_i}\left(\rho u_i T\right) = \frac{\partial}{\partial x_i}\left(\frac{k}{C_p}\frac{\partial u_j}{\partial x_i}\right)$

General transport equation (for scalars):

$\frac{\partial}{\partial x_i}\left(\rho u_i \varnothing\right) = \frac{\partial}{\partial x_i}\left(\Gamma_\varnothing \frac{\partial \varnothing}{\partial x_i}\right) - S_\varnothing$

Fig. 2.9 The variation of Q^* with χ for bundle using FG-1a with isosceles, rectangular and elliptical ducts [6]

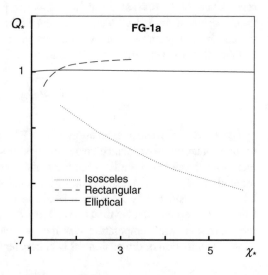

Table 2.6   Data for performance evaluation criteria [6]

Fixed								Objective
Case	A^*	D^*	L^*	W^*	P^*	Q^*	ϑ_i^*	
FG-1a	1	1	1	1	–	–	1	$Q^* > 1$
VG-1	–	1	–	1	1	1	1	$A^* < 1$
VG-2a	1	1	–	1	1	–	1	$Q^* > 1$

Fig. 2.10 Comparison of performance efficiencies of bundles with rectangular and hexagonal ducts, Case FG-a [6]

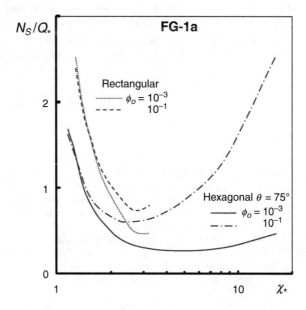

Fig. 2.11 Comparison of performance efficiencies of bundle with rectangular and hexagonal duct, Case VG-2a [6]

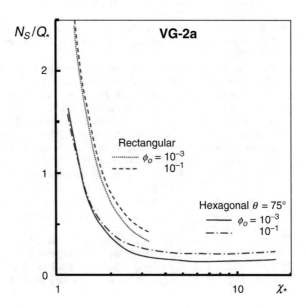

for different geometry of ducts. Table 2.6 shows the performance evaluation criteria with objective functions. Figure 2.10 shows the comparison of performance efficiencies of bundles with rectangular and hexagonal ducts, Case FG-1a. Figure 2.11 shows comparison of performance efficiencies of bundles with rectangular and hexagonal ducts, Case VG-2a.

The other information on PEC for single-phase flow has been presented by White and Wilkie [7], Usui et al. [8], Tauscher and Mayinger [9], Song and Gu [10], Sekulic and Kmecko [11], Sara et al. [12, 13], Raju and Bansal [14], Picon-Nunez et al. [15], Nunner [16], Norris [17], Miyashita et al. [18], Manzoor et al. [19], London [20], Le Foll [21], Kumada [22], Kreith and Black [23], Kern and Kraus [24] and Horvath [25].

References

1. Bergles AE, Bunn RL, Junkhan GH (1974) Extended performance evaluation criteria for enhanced heat transfer surfaces. Lett Heat Mass Transfer 1:113–120
2. Garcia A, Vicente PG, Viedma A (2005) Experimental study of heat transfer enhancement with wire coil inserts in laminar-transition-turbulent regimes at different Prandtl numbers. Int J Heat Mass Transfer 48(21–22):4640–4651
3. Vicente PG, Garcia A, Viedma A (2002) Heat transfer and pressure drop for low Reynolds turbulent flow in helically dimpled tubes. Int J Heat Mass Transfer 45(3):543–553
4. Bergles AE (1981) Applications of heat transfer augmentation. In: Kakac S, Bergles AE, Mayinger F (eds) Heat exchangers: thermal hydraulic fundamentals and design. Hemisphere, Washington, DC
5. Gholami A, Wahid MA, Mohammed HA (2017) Thermal–hydraulic performance of fin-and-oval tube compact heat exchangers with innovative design of corrugated fin patterns. Int J Heat Mass Transfer 106:573–592
6. Petkov VM, Zimparov VD, Bergles AE (2014) Performance evaluation of ducts with non-circular shapes: laminar fully developed flow and constant wall temperature. Int J Therm Sci 79:220–228
7. White WJ, Wilkie L (1970) The effect of rib profile on heat transfer and pressure loss properties of transversely ribbed roughened surfaces. In: Bergles AE, Webb RL (eds) Augmentation of convective heat and mass transfer. ASME, New York, pp 44–54
8. Usui H, Sano Y, Iwashita K, Isozaki A (1986) Enhancement of heat transfer by a combination of internally grooved rough tube and a twisted tape. Int Chem Eng 26(1):97–104
9. Tauscher R, Mayinger F (1998) Heat transfer enhancement in a plate heat exchanger with rib-roughened surfaces. In: Kakaç S (ed) Energy conservation through heat transfer enhancement of heat exchangers. Nato Advanced Study Institute, Cesme, İzmir, Turkey, pp 121–135
10. Song D, Gu W (1990) The optimization analysis calculation for high performance heat exchanger. In: Deng SJ, Veziroğlu TN, Tan YK, Chen LQ (eds) Heat transfer enhancement and energy conservation. Hemisphere, New York, pp 535–542
11. Sekulic DP, Kmecko I (1995) Three-fluid heat exchanger effectiveness-revisited. J Heat Transfer 117:226–229
12. Sara ON, Pekdemir T, Yapıcı S, Yılmaz M (2001a) Enhancement of heat transfer from a flat surface in a channel flow by attachment of rectangular blocks. Int J Energy Res 25(7):563–576
13. Sara ON, Pekdemir T, Yapici S, Yilmaz M (2001b) Heat-transfer enhancement in a channel flow with perforated rectangular blocks. Int J Heat Fluid Flow 22:509–518

14. Raju KSN, Bansal JC (1981) Design of plate heat exchangers, in low Reynolds number forced convection in channels and bundles. In: ASI proceedings, Ankara, Turkey, pp 597–616
15. Picon-Nunez M, Polley GT, Tores-Reyes E, Gallegos-Munoz A (1999) Surface selection and design of plate-fin heat exchangers. Appl Therm Eng 19:917–931
16. Nunner W (1958) Heat transfer and pressure drop in rough pipes. AERE Lib/Trans, p 786
17. Norris RH (1939) Proceedings of the fifth international congress of applied mechanics, p 585
18. Miyashita H, Fukushima K, Kometani M, Yamaguchi S (1990) Enhanced heat transfer mechanism using turbulence promoters in rectangular duct. In: Deng SJ, Veziroğlu TN, Tan YK, Chen LQ (eds) Heat transfer enhancement and energy conservation. Hemisphere, New York, pp 159–166
19. Manzoor M, Ingham DB, Heggs PJ (1983) The one-dimensional analysis of fin assembly heat transfer. J Heat Transfer 105:646–651
20. London AL (1964) Compact heat exchangers: Part 2. surface geometry. Mech Eng 86:31–34
21. Le Foll J (1957) Experimental research heat transfer. La Houille Blanche 1:30–45
22. Kumada M (1998) A study on the high performance ceramic heat exchanger for ultra high temperatures. In: Kakaç S (ed) Energy conservation through heat transfer enhancement of heat exchangers. Nato Advanced Study Institute, Cesme, İzmir, Turkey, pp 597–620
23. Kreith F, Black WZ (1980) Basic heat transfer. Harper & Row, New York
24. Kern DQ, Kraus AD (1972) Extended surface heat transfer. McGraw-Hill, New York
25. Horvath CD (1977) Three-fluid heat exchangers of two and three surfaces. Period Polytech 1:33–44

Chapter 3
Performance Evaluation Criteria Based on Laws of Thermodynamics

3.1 Variants in PEC

The PEC may involve fixed pressure drop instead of fixed pumping power. Balance point on the fan or pump curve may also be another criteria ([1] and Fig. 3.1(P67)). The system resistance is the sum of the friction loss in the condenser tubes and the piping and fittings external to the condenser. System characteristic curves may be the polynomial functions chosen by the designer when friction characteristics of the external piping and the condenser tubes are known.

Most of the PECs consider that total thermal resistance is on one side of the heat exchanger [2]. Cowell [3] compared compact heat exchanger surfaces. The hydraulic diameter may or may not vary.

Goodness factor comparison as a PEC has been used by Kays and London [4] and Shah [2]. Following equations for the PEC are to be considered:

$$h\eta_o = \frac{C_p \mu}{Pr^{\frac{2}{3}}} \frac{\eta_o jRe}{D_h} \qquad (3.1)$$

$$\frac{P}{A} = \frac{\mu^3}{2\rho^2} \frac{fRe^3}{D_h^{\,3}} \qquad (3.2)$$

$$\frac{h\eta_o A}{V} = \frac{4\sigma C_p}{Pr^{\frac{2}{3}}} \cdot \frac{jG}{D_h} \qquad (3.3)$$

$$\frac{P}{V} = \frac{2\sigma fG^3}{\rho^2 D_h} \qquad (3.4)$$

Fig. 3.1 Illustration of the
effect of two-phase pressure
drop on the LMTD

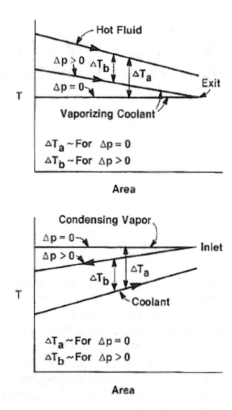

Soland et al. [5, 6], Fig. 3.2, have studied air-side performance of 17 parallel plate-fin surface geometries given in Kays and London [4]. Although this indicates compactness, no idea about the relative heat transfer surface area or fin material requirements is obtained from their method.

Yilmaz et al. [7] presented the PEC based on first law of thermodynamics for heat transfer enhancement in heat exchangers. A review of about 100 criteria has been presented with their characteristics and different constraints. Also, the relation between some of the criteria has been established and presented. There are simple dimensionless numbers like Nusselt number, j factor and friction factor which may be used to evaluate the performance of heat transfer enhancement techniques. But the evaluation is ineffective until it is expressed in terms of more than two factors like heat transfer rate, pumping power requirement of the fluid and geometry of the tube and inserts.

Another classification of performance evaluation criteria has been presented by the researchers. They subdivided the PEC into seven groups:

1. Comparison of heat exchanger surface keeping heat flux, mass flow rate of working fluid and hydraulic losses constant
2. Comparison of heat power maintaining constant geometry, flow rates and hydraulic losses

Fig. 3.2 V/V_s, for constant hA and P using Case VG-1 [5]

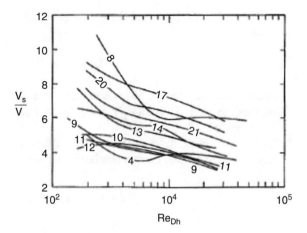

Plate-and-Fin Surfaces Compared

Surface geometry	Plate spacing, b (mm)	Surface code	Surface number
Plain	6.4	11.1	4
	6.4	19.86	8
Louvered	6.4	3/8–6.06	9
	6.4	3/8(a)–6.06	10
	6.4	1/2–6.06	11
	6.4	1/2(a)–6.06	12
	6.4	3/8–8.7	13
	6.4	3/8(a)–8.7	14
	6.4	1/4(b)–11.1	17
	6.4	1/2–11.1	20
	6.4	3/4–11.1	21

3. Comparison of hydraulic losses with mass flow rate of the fluid, geometry and heat power being constant
4. Evaluation based on economic accounting
5. Evaluation based on effective energy point of view
6. Evaluation based on exergy or second law analysis
7. Evaluation based on fuzzy comprehensive criteria

There are two methods in which heat transfer coefficient can be correlated for single-phase flow.

1. According to the first method, Nusselt number (Nu) has been expressed as a function of Reynolds number (Re) and Prandtl number (Pr):

$$Nu = \frac{hd}{k} = CRe^m Pr^n \tag{3.5}$$

They have also presented the same equation in a slightly different manner as follows:

$$J_h = NuPr^{-n} = CRe^m \tag{3.6}$$

where J_h has been termed as the heat transfer factor. This heat transfer factor can be used to plot J_h versus Re for graphical representation of heat transfer augmentation.

2. On the other hand, the Stanton number (St) has been expressed as a function of Reynolds number and Prandtl number in the second method:

$$StPr^{1-n} = CRe^{m-1} = j \tag{3.7}$$

$$St = \frac{h}{\dot{m} c_p} = \frac{Nu}{RePr} \tag{3.8}$$

In the graphical representation of j versus Re, j gives the same trend as that of friction factor (f) in f versus Re and thus j and f can be compared easily.

From equation,

$$J_h = jRe \tag{3.9}$$

Reported that the performance characteristics of various heat transfer techniques in terms of St, Nu and f versus Re have certain drawbacks like:

(a) The effect of various geometrical parameters of tube and passive enhancement elements used may not be included.
(b) The correlations might not be applicable to multi-fluid flow.

The surface performance comparison methods have been shown in Table 3.1.

Shah and London [10], Kays and London [4] and Sundén [11] worked on direct comparison of j and f. j/f ratio is called flow area goodness factor by Shah [12]. The frontal area of the heat exchanger which is required for a given heat transfer rate and pressure drop is indicated by the j/f ratio. Higher j/f denotes lesser frontal area requirement.

Kakaç [13] worked on comparison of heat transfer as a function of fluid pumping power. Kakac and Liu studied PEC based on pumping power per unit heat transfer area. Kays and London [14] and Soland et al. [6] studied pumping power per unit volume-based PEC. Figure 3.3 shows performance curves for different criteria. Figure 3.3a, d, c and d shows the performance curves for (a) same shape and volume of heat exchanger case, (b) same exchanger volume and pumping power case, (c) same pumping power and NTU case and (d) same volume and NTU case, respectively.

McClintock was the first to present correlations for performance evaluation criteria based on the rate of irreversibility for passage of the fluid on both sides of

Table 3.1 Classification of surface performance comparison methods [7]

Surface performance comparison methods			
Direct comparison of j and f	Comparison of heat transfer as a function of fluid pumping power	Performance comparison with a reference surface	Miscellaneous direct comparison methods
Flow area goodness factor	Pumping power per unit heat transfer area	Enhancement ratio	The ratio j/j_s and f/f_s
	Pumping power per unit volume	Bergles' PEC [8]	Efficiency index $\eta_c - (j/j_s)/(f/f_s)$
	Ratio of heat transfer to pumping power	Webb–Eckert's PEC [9]	Nu/Nu_s versus f/f_s at constant flow rate
		Cowell's PEC [3]	Thermal performance ratio
			Parameter $(f\,Re^2/j)$
			Heat transfer efficiency
			PEC for extended surfaces Fin effectiveness, ε_f
			$F_f n$ efficiency, η_f
			Overall surface efficiency, η_o
			Augmentation factor, AUG
			Enhancement factor, ζ

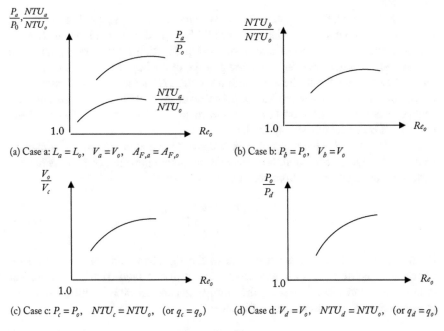

(a) Case a: $L_a = L_o$, $V_a = V_o$, $A_{F,a} = A_{F,o}$

(b) Case b: $P_b = P_o$, $V_b = V_o$

(c) Case c: $P_c = P_o$, $NTU_c = NTU_o$, (or $q_c = q_o$)

(d) Case d: $V_d = V_o$, $NTU_d = NTU_o$, (or $q_d = q_o$)

Fig. 3.3 Typical performance comparison results [6]

Fig. 3.4 Schematic diagram of heat transfer process in a single-fluid heat exchanger with prescribed heat flux distribution [15]

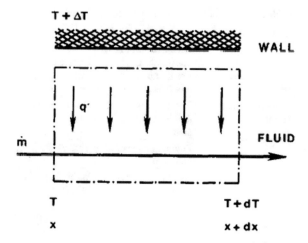

a heat exchanger. Bejan [15] presented a general criterion which can be used for evaluating the performance of a heat exchanger. He described two kinds of losses associated with heat transfer process in a heat exchanger: (a) losses due to fluid-to-fluid temperature difference, termed as ΔT losses, and (b) frictional losses also called ΔP losses. The ΔT losses can be reduced by increasing the surface area. But the increased surface results in increased ΔP losses. Thus, Bejan [15] used rate of irreversibility as the criterion to evaluate the heat transfer performance of a heat exchanger.

For this purpose, the number of entropy production units (N_s) has been proposed for the first time which was later used for all PEC based on second law analysis. It can be defined as the ratio of rate of entropy production (irreversibility rate) and stream-to-stream heat transfer rate to the passage of the heat exchanger. As $N_s \rightarrow 0$, the heat exchanger should be ideal with no losses (ΔT losses and ΔP losses). The schematic of heat transfer process in a heat exchanger with single fluid and pre-scribed heat flux distribution has been shown in Fig. 3.4. The analysis has been carried out based on this simple model.

Bejan [16, 17] give the exergy-based or entropy generation minimisation method from thermodynamic second law analysis to determine the merit of enhanced surfaces. The relevant equations for this are as follows:

$$S_{gen} = \frac{q' \Delta T}{T^2} + \frac{W}{\rho \Delta T}\left(-\frac{dP}{dx}\right)$$ (3.10)

From the above equation, the first term of the right-hand side of the expression represents the rate of entropy generation caused due to fluid friction. The second term of the expression represents the entropy generation rate resulting from heat transfer across finite temperature difference:

$$S_{gen} = S_{gen,T}(1 + \phi) \tag{3.11}$$

$$\text{where} \quad \phi = \frac{S_{gen,f}}{S_{gen,T}} \tag{3.12}$$

S_{gen} being the entropy generation rate per unit length of heat exchanger passage,

$$N_s = \frac{S_{gen}}{S_{gen,s}} \tag{3.13}$$

where N_s is the entropy generation number which quantifies the enhancement.
Also,

$$N_s = \frac{N_T + \phi_s N_P}{1 + \phi_s} \tag{3.14}$$

$$N_T = \left(\frac{St_s}{St}\right)\left(\frac{D_h}{D_{h,s}}\right) \tag{3.15}$$

$$Re = \left(\frac{D_h}{D_{h,s}}\right)\left(\frac{A_s}{A}\right)Re_s \tag{3.16}$$

$$\phi_s = \left(\frac{T}{\Delta T}\right)_s^2 \left(\frac{G^2}{A_c C_p T}\right)_s \left(\frac{f_s/2}{St_s}\right) \tag{3.17}$$

where ϕ_s is the irreversibility distribution ratio of the smooth surface.

The equation for N_s (number of entropy production units) according to Bejan [15] is as follows:

$$N_s = \frac{T}{q'}\frac{d\dot{s}}{dx} = \frac{\dot{m}}{\rho q'}\left(-\frac{dP}{dx}\right) + \frac{\Delta T}{T}\left(1 + \frac{\Delta T}{T}\right)^{-1} \tag{3.18}$$

The effect of wall-fluid temperature difference ΔT and a combined parameter A (which considers the dependence of Re along with duty parameter dependence on N_s) has been shown in Fig. 3.5. Also, the effect of A and ratio of heat transfer coefficient to pumping power required (represented by R) has been shown in Fig. 3.6. They have concluded that the number of entropy generation units (N_s) criterion serves well to evaluate the performance of heat exchangers.

Bejan and Pfister Jr [18] studied different augmented surfaces in order to achieve reduced irreversibility or energy destruction. They evaluated the performance of heat exchangers by comparing the entropy generation rate in enhanced channel with that in a smooth channel which has been denoted by N_s. They studied the performance of tubes with sand grain roughness and repeated rib roughness. The variation of entropy

Fig. 3.5 The local number
of entropy production units
N, as a function of the wall
fluid ΔT and the combined
parameter A [15]

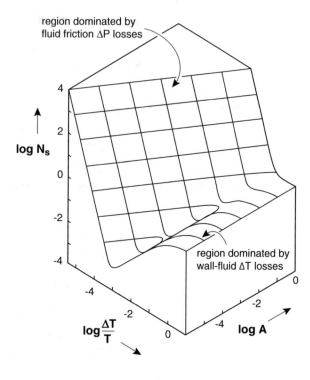

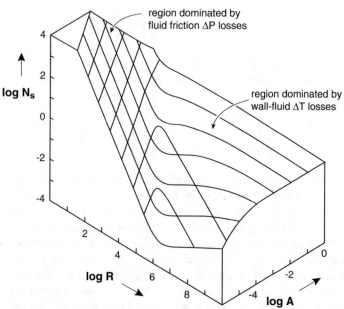

Fig. 3.6 The local number of entropy production units N_s as a function of A and the ratio of heat
transfer coefficient to fluid pumping power R [15]

generation number (N_s) has been shown for sand roughness and repeated rib roughness in Figs. 3.7 and 3.8, respectively. Also, the optimum rib height-to-tube diameter ratio (e/D) for minimum energy destruction rate has been shown in Fig. 3.9M (dimensionless pitch, $P/e = 10$). They conclude that the roughness techniques enhanced the heat transfer rate, but did not result in reduced

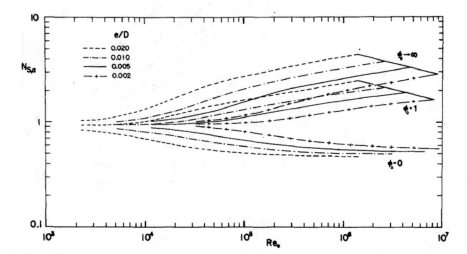

Fig. 3.7 Augmented entropy generation number for sand grain roughness [18]

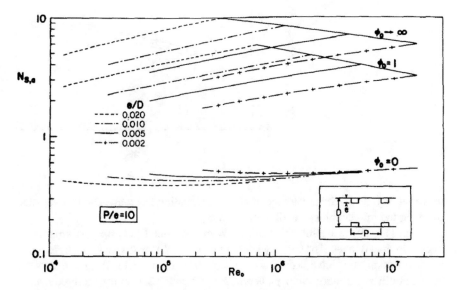

Fig. 3.8 Augmented entropy generation number for repeated-rib roughness [18]

Fig. 3.9 Optimum rib height for minimum exergy destruction [18]

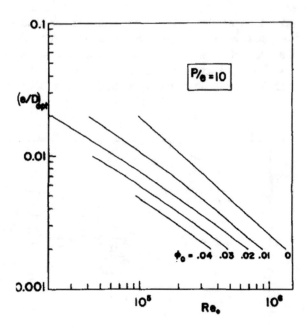

Fig. 3.10 Entropy generation number W_s associated with using n straight fins on the inside surface of a smooth tube [19, 20]

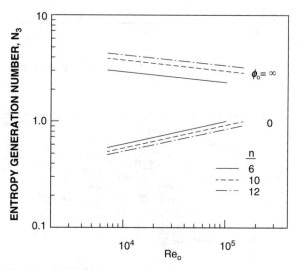

irreversibility. Thus, the advantage of any augmentation technique can be evaluated by using the irreversibility distribution ratio φ_o.

Bejan [19, 20] discussed the work of Webb and Scott [21] on performance of internally finned tubes. The entropy generation units (N_s) versus Re for n straight fins inside a smooth tube have been shown in Fig. 3.10. Figure 3.11 [22] represents the irreversibility distribution ratio φ_o versus Re plot which shows the thermodynamic feasibility of using fins to enhance heat transfer augmentation in smooth tubes.

Fig. 3.11 The irreversibility distribution ratio vs. Reynolds number [19, 20]

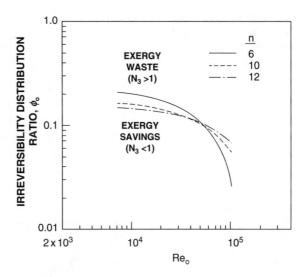

More details may be obtained from Nikuradse [23], Dipprey and Sabersky [24], Karman-Nikuradse data that may be obtained from Kays and Perkins [25], Chen and Huang [26], Zimparov and Vulchanov [27] and Zimparov [28–30].

Nag and Kumar [31] investigated and modified Bejan's entropy generation criterion by considering the variation of fluid temperature along heat transfer passage. Prasad and Shen [32] proposed a conclusive method which was based on exergy analysis. They introduced heat transfer improvement number (N_H) and exergy destruction number (N_E) for comparison of the effect of improved heat transfer with increased irreversibility.

Zimparov et al. [33] have presented a critical review on different performance evaluation criteria (PEC) used to evaluate heat transfer enhancement performance of various heat transfer augmentation techniques. They reported two types of evaluation methods based on first law of thermodynamics and second law of thermodynamics. The PEC based on the first law analysis and second law analysis has been studied by Yilmaz et al. [7] and Yilmaz et al. [34], respectively. The various objectives of performance evaluation criteria are to reduce surface area of heat transfer in order to obtain compact heat exchangers, to increase the rate of heat transfer and to reduce pumping power consumption. The variation on Nusselt number ratio and friction factor ratio (for augmented tube and smooth tube) with Reynolds number has been shown in Fig. 3.12 and Fig. 3.13, respectively. Zimparov et al. [33] discussed the performance evaluation criteria with increased heat transfer rate as the objective. Table 3.2 shows the comparison of different performance evaluation criteria.

Zimparov [28] experimentally studied ten tubes with spiral corrugations and presented extended performance evaluation criteria correlations for enhanced surfaces. The analysis was based on second law of thermodynamics and the effect of variation of fluid temperature along the length of the tube in axial direction has also

Fig. 3.12 Variation of the ratio of friction factors with Reynolds number [33]

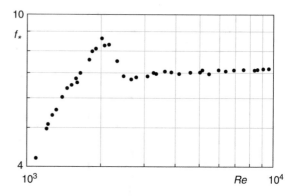

Fig. 3.13 Variation of the ratio of Nusselt numbers with Reynolds number for different *Pr* values [33]

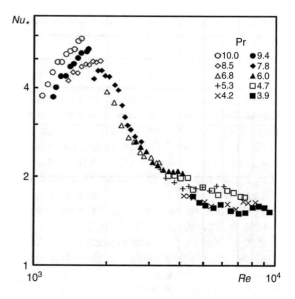

been taken into account. The experiments were carried out under constant wall temperature boundary condition. The geometry of the tubes has been presented in Table 3.3. The variation of St and F with Re for smooth tube and rough tube has been shown in Fig. 3.14. The effect of increased heat rate, augmented entropy generation number, reduced tube length and reduced surface area has been discussed. The results indicated that the ratio of rib height and tube diameter (e/D) was optimum at 0.04 for the considered spirally corrugated tubes. The performance of enhanced surfaces at both constant wall heat flux and constant wall temperature boundary conditions has been studied. The comparison has been presented in Fig. 3.15. It has been concluded that better performance of the enhanced tubes was observed for constant wall temperature boundary condition.

The performance evaluation of three-start spirally corrugated tubes in combination with twisted tapes has been studied by Zimparov [29]. The details of the

Table 3.2 Benefits evaluated by using criteria $R_{3,b}$, $R_{3,a}$ and PEC [33]

Re_a	Re_s	$R_{3,b}$	$R_{3,a}$	PEC
Pr = 10				
1080	2170	2.00	2.42	2.27
1200	2390	1.89	2.88	2.70
1360	2620	1.86	3.21	2.99
1530	2980	1.84	3.39	3.13
Pr = 9.5				
1220	2340	1.90	2.36	2.21
1440	2690	1.87	2.72	2.51
1600	2980	1.83	3.03	2.80
1830	3420	1.70	2.92	2.68
Pr = 8.5				
1390	2660	1.78	2.52	2.33
1640	3140	1.67	2.76	2.54
1880	3580	1.56	2.78	2.56
2060	3960	1.56	2.74	2.51
Pr = 7.8				
1780	3440	1.47	2.46	2.27
1840	3580	1.50	2.55	2.35
2040	3970	1.48	2.35	2.15
2370	4610	1.41	1.80	1.65
2770	5460	1.36	1.55	1.44
Pr = 6.8				
2000	3960	1.30	2.04	1.85
2450	4900	1.29	1.55	1.40
2840	5680	1.26	1.33	1.22
3290	6590	1.23	1.29	1.19
Pr = 6.0				
2640	5320	1.25	1.51	1.39
3090	6240	1.17	1.25	1.15
3750	7600	1.15	1.19	1.11
4250	8560	1.14	1.16	1.07
Pr = 5.3				
3320	6770	1.09	1.19	1.10
4350	8820	1.04	1.05	0.96
5680	11,470	1.05	1.06	0.98
6820	13,800	1.06	1.06	0.97

geometry of the tubes used in the study have been shown in Table 3.4. The increase in heat transfer coefficients for the enhanced tubes has been shown in Table 3.5. The ratio of $\frac{N_s}{Q_*}$ has been used to evaluate the performance of enhanced tubes because it takes both the first law analysis and second law analysis into account. Thus, the variation of $\frac{N_s}{Q_*}$ with Re has been shown in Figs. 3.16, 3.17 and 3.18 based on FG1a, FG2a and VG2a, respectively.

Table 3.3 Tube geometry [28]

Tube no.		Reference	D_o (mm)	D_i (mm)	e (mm)	p (mm)	e/D	p/e	β_*	E_o
1	2a		25.40	23.57	0.271	2.54	0.012	9.4	0.935	1.64
2	2b	[3]	25.40	23.57	0.393	6.35	0.017	16.1	0.840	1.29
3	2c		25.40	23.57	0.751	12.70	0.032	16.9	0.698	1.09
4	6		25.30	23.39	0.886	9.75	0.038	11.0	0.906	1.07
5	7	[79]	25.30	23.42	0.775	9.40	0.033	12.1	0.919	1.10
6	2200		24.61	21.90	0.439	6.35	0.020	14.5	0.941	1.25
7	14		25.38	22.00	0.947	13.46	0.043	14.2	0.878	1.04
8	18		27.27	25.20	1.022	10.95	0.041	10.7	0.919	1.07
9	33	[51]	27.56	25.62	0.447	6.55	0.017	14.7	0.952	1.20
10	34		27.62	25.78	0.628	8.48	0.024	13.5	0.938	1.08

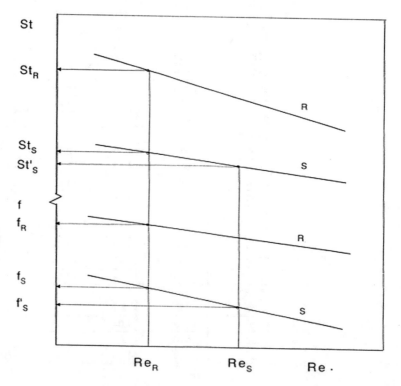

Fig. 3.14 The behaviour of the Stanton number and friction factor vs. Reynolds number [28]

Zimparov [30] presented many plots between heat transfer rate and augmentation entropy versus Reynolds number for fixed geometry (FG) and variable geometry (VG-1). The ratio of N_s/Q^* versus Reynolds number, group $N_S A^*$ versus Reynolds number and defined correlations for Nu (St) and friction factor f of the augmented

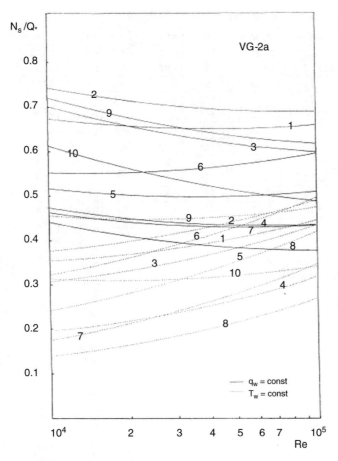

Fig. 3.15 The ratio $N_s = Q^*$ vs. the Reynolds number. Comparison between the boundary conditions $q_w =$ constant and $T_w =$ constant [28]

surface as a function of Reynolds number (Re) have also been plotted. Zimparov analysed the first law of thermodynamics and developed extended performance evaluation criteria equation which included the effect of fluid temperature variation. The newly developed performance evaluation criteria (PEC) for enhanced heat transfer surfaces consider the entropy generation and exergy destruction. Also, he observed that general evaluation criterion enriches the Bejan's EGM method. This method serves in recognising the unsuitable enhanced surface and directs the designer in proper direction.

Ouellette and Bejan [22] used second law analysis to evaluate the performance of various swirl promoters. The schematic representation of the inserts has been shown in Fig. 3.19. The variation of number of entropy generation units ($N_{s,a}$) for augmented surfaces [internal fins (IF), internal spiral fins (ISF), twisted tapes (TT) and helical tube (H)] with Reynolds number has been shown in Fig. 3.20. They

Table 3.4 Characteristic parameters of the tubes [29, 30]

Number	D_o (mm)	D_i (mm)	e (mm)	p (mm)	β (°)	e/D_i	p/e	β_*	H/D_i
4040	15.72	13.68	0.557	5.97	67.4	0.0407	10.73	0.749	–
4041	15.72	13.68	0.557	5.97	67.4	0.0407	10.73	0.749	15.4
4042	15.72	13.68	0.557	5.97	67.4	0.0407	10.73	0.749	12.3
4043	15.72	13.68	0.557	5.97	67.4	0.0407	10.73	0.749	7.8
4044	15.72	13.68	0.557	5.97	67.4	0.0407	10.73	0.749	5.9
4045	15.72	13.68	0.557	5.97	67.4	0.0407	10.73	0.749	4.8
4030	15.72	13.73	0.781	5.82	68.0	0.0569	7.45	0.755	–
4031	15.72	13.73	0.781	5.82	68.0	0.0569	7.45	0.755	15.3
4032	15.72	13.73	0.781	5.82	68.0	0.0569	7.45	0.755	12.2
4033	15.72	13.73	0.781	5.82	68.0	0.0569	7.45	0.755	7.7
4034	15.72	13.73	0.781	5.82	68.0	0.0569	7.45	0.755	5.8
4035	15.72	13.73	0.781	5.82	68.0	0.0569	7.45	0.755	4.7

Table 3.5 Increase of the heat transfer coefficient [29, 30]

Number	Nu_R/Nu_S
4040	1.90
4041	2.3–2.4
4042	2.3–2.6
4043	2.5–3.5–3.1
4044	2.7–4.5–3.7
4045	2.8–5.5–4.2
4030	2.3–2.2
4031	2.7–3.4–3.0
4032	2.7–3.9–3.4
4033	3.0–5.0–4.6
4034	3.6–7.3–6.3
4035	3.6–9.6–7.4

concluded that $N_{s,a}$ for twisted tape and helical tube are minimum and thus they are preferred over internal straight and spiral fins.

Zimparov et al. [35] reported the advantage of compound techniques for single-phase turbulent flow in shell and tube-type heat exchangers. The performance of combined technique of using surface roughness with twisted tape inserts has been studied. The PEC based on both first law and second law of thermodynamics has been used. The following factors have been considered in carrying out the study:

- The thermal resistance which is external to the surface has been taken into account by considering the outside heat transfer coefficient.
- The fouling resistances on the two sides of the wall have been neglected.
- The performance has been evaluated at constant wall temperature boundary condition.

The performance has been evaluated for two cases, namely, fixed geometry criteria (FG) and variable geometry criteria (VG). The fixed geometry criteria

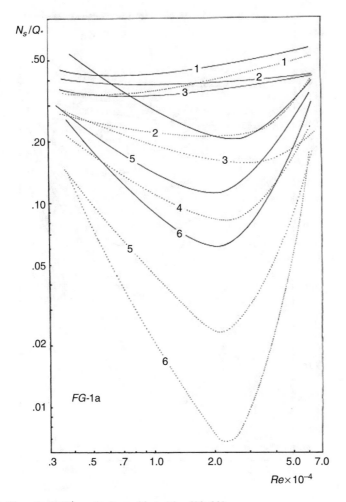

Fig. 3.16 The ratio N_s/Q^* vs. the Reynolds number [29, 30]

consider the replacement of smooth tubes with enhanced surfaces having the same basic geometry. On the other hand, variable geometry criteria are used for designing heat exchangers for a required thermal duty for a given flow rate of the fluid. In order to accommodate large pressure drop characteristics of the tube with compound passive inserts, the tube-side velocity of the flow has to be reduced by varying frontal area of the flow to meet the pumping power requirement. They have used 17 test tubes for the study and their geometrical parameters were shown in Tables 3.6 and 3.7.

The ratio of enhancement entropy generation (N_s) and non-dimensional heat transfer rate (Q_*) has been used for performance evaluation. This ratio $f(Re_R) = \frac{N_s}{Q_*}$ relates the first law analysis and second law analysis. The non-dimensional heat

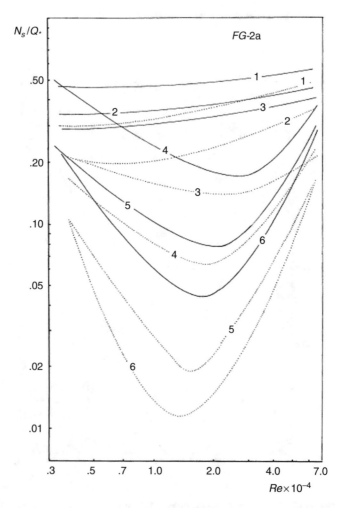

Fig. 3.17 The ratio N_S/Q^* vs. the Reynolds number [29, 30]

transfer rate is the ratio of heat transfer rates in rough tube and smooth tube. The variation of $\frac{N_s}{Q_*}$ with Reynolds number for all the 17 tubes considered in their study has been shown in Figs. 3.21 and 3.22. Their results showed that tubes 1–3 were ineffective as compared to the smooth tube. The tube 5 was observed to be beneficial for FG case and higher Reynolds numbers. Tube 11 was observed to perform well for all values of Reynolds number. The internal 3D extended surfaces in combination with twisted tape inserts were found suitable for laminar flow rather than turbulent flow.

Bishara et al. [36] numerically evaluated the effect of swirl flow on the performance of heat transfer enhancement in axially twisted oval tubes with elliptical cross sections. The numerical analysis was done on water with some constant properties

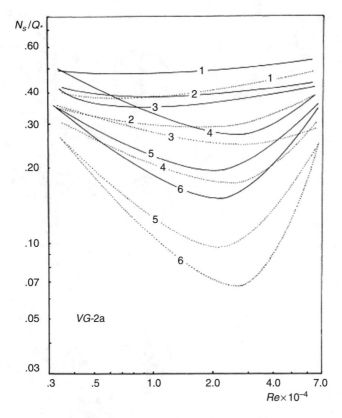

Fig. 3.18 The ratio N_s/Q^* vs. the Reynolds number for VG-2a [29, 30]

such as Prandtl number ($Pr \sim 3.0$), Reynolds number ($10 \le Re \le 1000$), twist ratio for 180° rotation of twisted tube ($3.0 \le y \le 6.0$) and aspect ratio ($0.3 \le \alpha \le 0.7$). They also studied the effect of swirl on the velocity distribution, isothermal fanning factor and Nusselt number for a tube maintained at uniform wall temperature. They observed 2.5 times higher heat transfer rate relative to an equivalent straight tube for constant pumping power. Shah and Sekulic [37], Shah and London [10] and Kays and London [38] gave two commonly used enhancement evaluation criteria for the design of compact heat exchanger: (1) the area goodness factor (j/f), where j ($=Nu/Re\,Pr^{1/3}$), and (2) the volume goodness factor. Figure 3.23 suggests the basis of area goodness factor that twisted tubes are not much effective in reducing the frontal area of the tube bundle. Both increased heat transfer rate and pumping power due to higher frictional loss were analysed by plotting the graph between (jRe) and (fRe^3). Figure 3.24 presents the thermal-hydrodynamic performance of two different oval tubes having aspect ratios 0.3 and 0.7 with twist ratios 3.0 and 6.0 on the basis of volume goodness factor.

Zimparov and Vulchanov [27] studied the performance evaluation criteria and presented equations based on the entropy generation theorem. They have considered

Fig. 3.19 Schematic of four
swirl flow-promoting
techniques for heat transfer
augmentation [22]

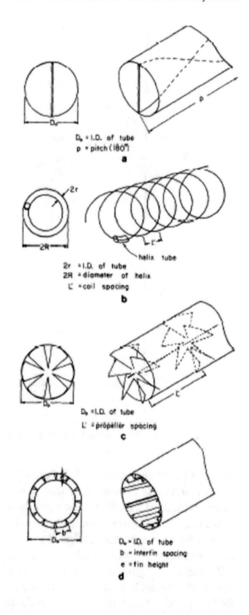

the fixed geometry criteria (FG) for their analysis. This criterion involves replacement of smooth tube by augmented tube one by one, respectively. It considers increased heat duty or UA for a constant flow rate and velocity of the heat exchangers. The constraints $W^* = 1$, $N^* = 1$ and $L^* = 1$ require $Re_s = D^* Re_R$ and $P^* > 1$. Surface promoters are one of the best augmentation techniques in which wall roughness has negligible impact on hydraulic diameter D_h and flow cross section. Thus, it can be assumed that $D^* = 1$.

Fig. 3.20 Comparison of four swirl promoters based on irreversibility minimisation potential; IF = internal straight fins; ISF = internal spiralled fins; TT = twisted tape; H = helical tube [22]

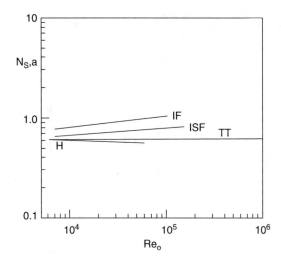

Table 3.6 Geometrical parameters of the tested tubes [35]

No.	D_i (mm)	e (mm)	p (mm)	β (°)	e/D_i	p/e	β_*	H/D_i	NDC
1	6.35	0.076	–	–	0.012	–	–	12.32	–
2	6.35	0.076	–	–	0.012	–	–	8.34	–
3	6.35	0.076	–	–	0.012	–	–	5.10	–
4	14.17	0.309	1.04	30.0	0.022	3.37	0.333	8.40	–
5	14.17	0.309	1.04	30.0	0.022	3.37	0.333	8.40	–
6	13.90	0.312	5.76	82.4	0.022	18.5	0.916	4.68	–
7	13.15	0.593	5.06	83.0	0.045	8.50	0.922	4.94	–
8	13.51	0.767	5.19	83.0	0.057	6.67	0.922	5.83	–
9	19.20	0.350	3.02	11.0	0.018	8.62	0.122	4.16	0.13
10	19.20	0.350	3.02	11.0	0.018	8.62	0.122	5.21	0.02
11	19.50	0.850	3.06	–	0.044	3.60	–	1.03	–
12	19.20	0.380	3.02	11.0	0.020	7.95	0.122	1.82	0.04
13	19.50	0.850	3.06	–	0.044	3.60	–	1.03	0.18
14	19.20	0.380	3.02	11.0	0.020	7.95	0.122	1.82	0.21
	e/D_i		p_a/e		w/p_a		p_c/w		H/D_i
15	0.077		4.080		0.118		5.421		30.0
16	0.077		4.080		0.118		5.421		20.0
17	0.077		4.080		0.118		5.421		10.0

Table 3.7 Geometrical parameters of the 3-DIEST tubes

	H (mm)	D_d (mm)	d (mm)	δ (mm)	Helical direction
Coil A	20.0	17.8	2.0	–	Clockwise
Coil B	35.0	17.8	2.0	–	Clockwise
Twisted tape A	80.0	16.2	–	1.0	Counterclockwise
Twisted tape B	100.0	18.2	–	1.0	Counterclockwise

Fig. 3.21 Variation of N_s/Q^* with Re [35]

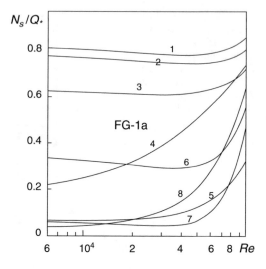

Fig. 3.22 Variation of N_s/Q^* with Re [35]

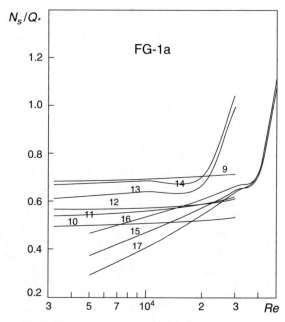

For increased heat duty $Q^* > 1$, the augmentation entropy generation number N_s is

$$N_s = \frac{1}{1 + \phi_o}\left(\frac{Nu_s}{Nu_R}Q^{*2} + \phi_o\left(\frac{f_R}{f_s}\right)\right) \qquad (3.19)$$

where $N_T = Q^{*2}\left(\frac{Nu_S}{Nu_R}\right)$ and $N_P = \left(\frac{f_R}{f_s}\right)$.

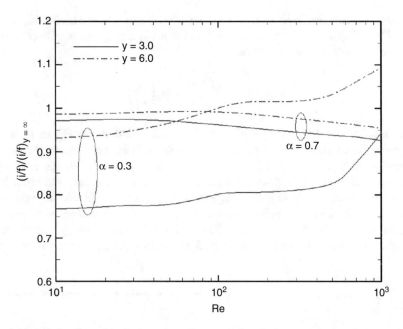

Fig. 3.23 Relative thermal-hydrodynamic performance of oval tubes, evaluated on the basis of the area goodness factor [36]

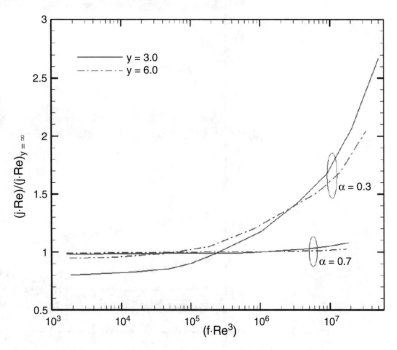

Fig. 3.24 Relative thermal-hydrodynamic performance of oval tube evaluated on the basis of the volume goodness factor [36]

Also, the constraints $N^* = 1$, $L^* = 1$ and $W^* = 1$ and $Q^* = 1$ require $Re_R = Re_S$ and $P^* > 1$. The requirement is $\Delta T^*_m < 1$ and thus the augmentation entropy generation number N_s can be written as

$$N_s = \frac{1}{1 + \phi_o}\left(\frac{Nu_S}{Nu_R} + \phi_o\left(\frac{f_R}{f_S}\right)\right) \tag{3.20}$$

Zimparov and Vulchanov [27] plotted Figs. 3.25 and 3.26 for FG-1b in which augmentation technique was used to reduce the driving temperature difference. Similarly, they plotted Fig. 3.27 for FG-2a and Fig. 3.28 for FG-2b and concluded that smallest value of N_s does not confirm highest heat transfer rate. Figure 3.29 shows values corresponding to N_s in Fig. 3.28. Figure 3.30 for FG-2c represents extended case of FG-2, Fig. 3.31 represents FG-3 in which variation of P^* with Re for all tubes has been shown, Fig. 3.32 presents FN-1 case showing reduced tube length versus Reynolds number and Fig. 3.33 presents corresponding values of N_s.

Fig. 3.25 Augmentation entropy generation number vs. Reynolds number [27]

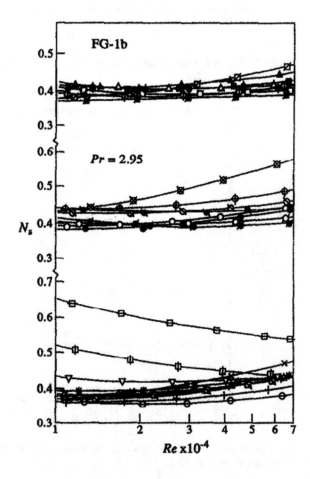

Fig. 3.26 Reduced driving temperature difference vs. Reynolds number [27]

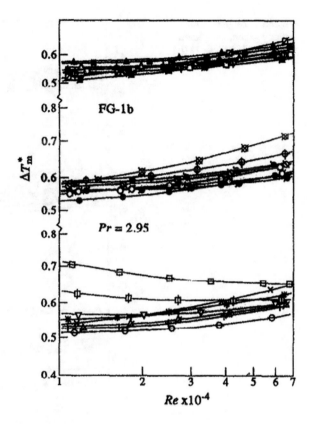

VG-1 case has been presented in Fig. 3.34 and VG-2a cases have been presented in Figs. 3.35 and 3.36.

These figures help in identifying the tube with better performance. They concluded from their investigation that evaluation of heat transfer enhancement techniques should be based on the first and second laws of thermodynamics for better performance. Their established equations enriched the PEC for augmented heat transfer surfaces which were developed by first law analysis in association with entropy generation and exergy destruction.

The works of Ahamed et al. [39], Lei et al. [40], Fan et al. [41], Guo et al. [42], Fan et al. [43], Zhang et al. [44], Xu et al. [45] and Esmaeilnejad et al. [46] are of great importance in the area of heat transfer enhancement but did not give the proper guidelines for performance evaluation. Lorenzini and Suzzi [47] studied the geometrical variation on thermal performance of fully developed steady laminar flow of a Newtonian liquid through microchannel in the presence of the viscous dissipation. The boundary conditions were uniform heat flux and uniform peripheral temperature across the cross section. The micro heat exchangers have small dimensions and permit very high heat transfer coefficients even in laminar flow. Also, they can be manufactured in several shapes. This makes micro heat exchangers attractive in many applications. Lorenzini [48] and Lorenzini and Morini [49] studied the

Fig. 3.27 Augmentation
entropy generation number
vs. Reynolds number [27]

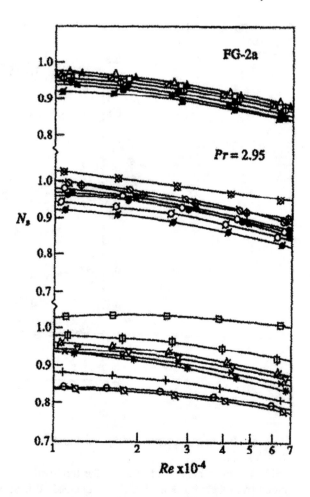

rounded corners in rectangular ducts. They analysed heat transfer and friction losses
and calculated *fRe* and *Nu* with and without viscous heating. Lorenzini and Suzzi
[47] took base geometry as rectangle with sides $2a$ and $2b$. They gradually rounded
the corners as shown in Fig. 3.37. The aspect ratio $\beta = 2a/2b$ has been varied from
$\beta = 0.1$, 0.50, 0.25, 0.10 and 0.03. Their objective function Q^* (non-dimensional
heat transfer to fluid) was as follows:

$$Q^* = \frac{Nu^* P_h^* L^* \Delta T^*}{D_h^*} \tag{3.21}$$

and non-dimensional pumping power is given by

$$P^* = \frac{m^{*2} (fRe)^* L^*}{A_C^* D_h^{*2}} \tag{3.22}$$

They modelled Brinkman number as

Fig. 3.28 Reduced driving
temperature
difference vs. Reynolds
number [27]

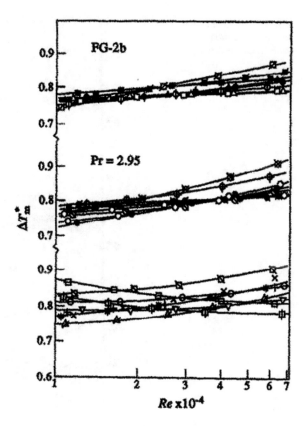

$$Br = \frac{Br_{\text{ref}}}{A_{\text{C}}^* P_{\text{h}}^{*2} Nu_{Br=0}^* (1 + \sigma_{\text{ref}} Br_{\text{ref}}) - \sigma Br_{\text{ref}}} \qquad (3.23)$$

This equation is the function of aspect ratio β and radius of curvature. The objective function Q^* versus radius of curvature has been plotted in Fig. 3.38 and entropy generation number for FG1a and $\beta = 1$ has been presented in Fig. 3.39. They considered various aspect ratios ($\beta = 0.1, 0.50, 0.25, 0.10$ and 0.03) with Brinkman number (10^{-1} to 10^{-3}) as reference. The reference augmentation irreversibility ratio is $\phi_{\text{ref}} = 10^{-1}$ unless or otherwise stated.

Lorenzini and Suzzi [47] used both first and second laws of thermodynamics and concluded that the Brinkman number related to objective function changes as it depends on the performance evaluation criteria and accounted constraints. Also, Q^* depends on the geometrical constraints. The best results of study were obtained for $P_{\text{h}}^* = 1$. They observed that non-dimensional heat transfer rate Q^* is affected by the geometrical constraints which was contrary to the no viscous dissipation case. Similar trends of N_s and Q^* were observed.

Chai et al. [50] computationally analysed the performance of microchannel heat sink in combination with fan-shaped ribs on sidewalls. They aimed at obtaining the

Fig. 3.29 Augmentation
entropy generation number
vs. Reynolds number [27]

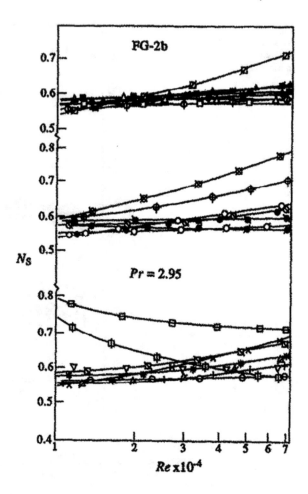

interrelation between thermal resistances and pumping power. They numerically calculated the entropy generation rate and performance evaluation criteria using water and silicon. They modelled fan-shaped ribs on two parallel sidewalls together in both aligned and offset arrangements as in Fig. 3.40. For clear and better understanding of mechanism and properties of microchannel heat sinks, they studied both the entropy generation rate due to heat transfer and fluid friction separately. They concluded that geometrical parameters of fan-shaped ribs have an impact on the thermal and hydraulic characteristics of microchannel heat sinks. The performance evaluation based on rib height has estimated that offset fan-shaped ribs were better than aligned-fan ribs. Also, the decrement in rib spacing resulted in gradual performance degradation. They found that large rib height and small rib spacing increase the total entropy generation rate. Thus, this leads to the performance degradation. They concluded that the performance of the tube with larger rib height and smaller rib spacing was worse than that of a smooth one. The best microchannel

Fig. 3.30 Augmentation entropy generation number vs. Reynolds number [27]

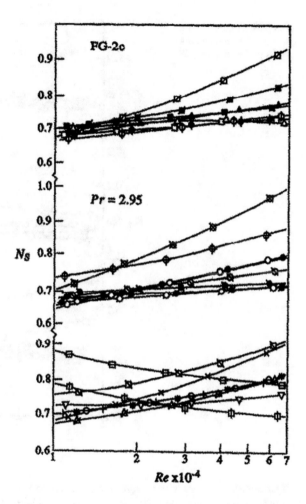

heat sinks showed 32% decrease in entropy generation rate and 1.33 in PEC with Reynolds number ranging from 187 to 715. They plotted the effects of W_r/S_r on PEC as shown in Fig. 3.41, effect of H_r/W_c on PEC as in Fig. 3.42 and effect of S_r/W_c on PEC as in Fig. 3.43 for result interpretations.

Dai et al. [51], Wang et al. [52, 53], Escandón et al. [54], Khan et al. [55], Koşar [56], Xia et al. [57, 58], Sharma et al. [59], Kandlikar et al. [60], and Shalchi-Tabrizi and Syef [61] worked on microchannel, minichannels and micro heat sink. Zhai et al. [62] studied the heat transfer performance of flow of de-ionised water. The de-ionised water flowing through complex-structured micro heat sinks in accordance with the uniform heat flux boundary condition has been studied. They proposed six types of cavities and ribs for micro heat sinks. These cavities and ribs were placed on the sidewall of microchannels along the direction of flow. The six cavities and rib combinations were (1) triangular cavities with circular rib (Tri.C-C.R in short),

Fig. 3.31 Augmentation
entropy generation number
vs. Reynolds number [27]

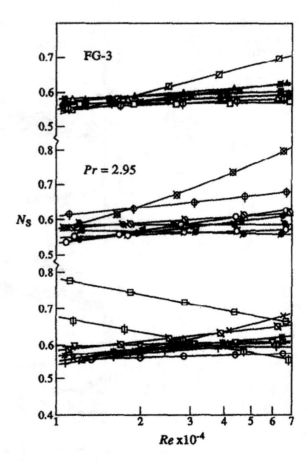

(2) triangular cavities with triangular rib (Tri.C-Tri.R in short), (3) triangular cavities with trapezoidal ribs (Tri.C-Tra.R in short), (4) trapezoidal cavities with circular rib (Tra.C-C.R in short), (5) trapezoidal cavities with triangular rib (Tra.C-Tri.R in short) and (6) trapezoidal cavities with trapezoidal rib (Tra.C-Tra.R in short). They are represented in Fig. 3.44.

Also, the schematic diagram of micro heat sink has been presented in Fig. 3.45. They analysed and concluded that the micro heat sink having triangular cavities with triangular ribs (Tri.C-Tri.R) performed well in the region of Reynolds number $300 < Re < 600$. They used Ansys FLUENT software and plotted Fig. 3.46 representing apparent friction factor with different ribs and cavities. Also, they compared Nusselt number and thermal enhancement factor for tubes with different cavities and ribs as shown in Figs. 3.47 and 3.48, respectively. The reasonable geometrical sizes have an impact on thermal management. They further analysed entropy generation and exergy destruction and found that heat transfer irreversibility decreases as the Reynolds number increases. The increase in Reynolds number resulted in increased irreversibility. Figure 3.49 presents the trend of entropy

Fig. 3.32 Reduced tubing length vs. Reynolds number [27]

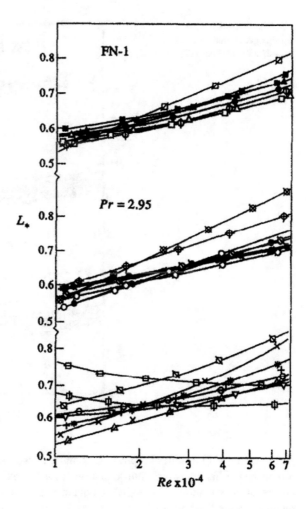

Fig. 3.32 Reduced tubing length vs. Reynolds number [27]

generation with Reynolds number variation. However, increase in Reynolds number leads to decrease in irreversibility. Also, they worked on development of correlation between friction and Nusselt number. It was presented as

$$Nu = 0.162Re^{0.7955}\left(\frac{e_1}{D_h}\right)^{-0.0376}\left(\frac{e_2}{D_h}\right)^{0.3501} \tag{3.24}$$

where $0\langle\frac{e_1}{D_h}\langle 0.7502, 0.12\langle\frac{e_2}{D_h}\langle 0.22$ and $260\langle Re \langle 600$. This predicts the values with $\pm10\%$ accuracy.

They plotted augmentation entropy generation and transport efficiency of thermal energy with relative cavity height, Fig. 3.50, and with relative rib height, Fig. 3.51. Also, they finally compared numerical friction factor with that predicted from correlation and has been presented in Fig. 3.52.

Fig. 3.33 Augmentation
entropy generation number
vs. Reynolds number [27]

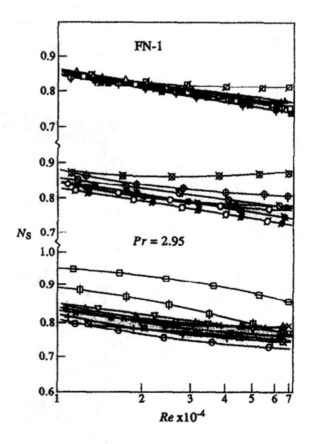

Liu et al. [63] introduced a physical quantity called enerty. They introduced and established different equilibrium equation from the existing conventional energy conservation equation. The existing equation associated with laminar flow for incompressible fluid can be represented as

$$\rho C_{\mathrm{P}} \frac{DT}{Dt} = \lambda \Delta^2 T + \phi + \dot{Q}''' \qquad (3.25)$$

Rigorous calculations and analysis by Liu et al. [63] had been done for hypothesising a physical quantity enerty which characterise energy both in quantity and in quality as

$$e = h - T_{\mathrm{o}} S \qquad (3.26)$$

The established equilibrium equation in terms of enerty is represented as

$$\rho \frac{D_e}{D_t} = -\nabla q - \frac{\lambda (\Delta T)^2}{T} + \phi + \dot{Q}^m \qquad (3.27)$$

Fig. 3.34 Augmentation
entropy generation number
vs. Reynolds number [27]

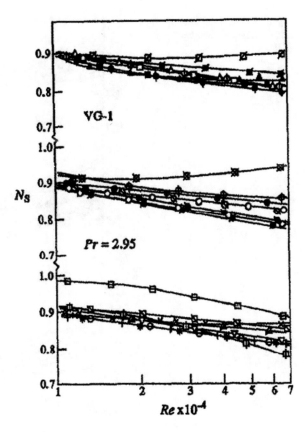

where $\frac{\lambda(\nabla T)^2}{T}$ is the inaccessible heat due to irreversible dissipation loss. Its unit is
energy per cube meter and may be termed as heat consumption of the fluid. The term
$\frac{\lambda(\nabla T)^2}{T}$ is always positive. It means that "The irreversible process always varies
towards the direction enerty decreasing due to the increase in heat consumption".
Further, Table 3.8 presents the relation between physical quantities like entropy,
enerty, entransy, entropy generation, heat consumption and entransy dissipation. Liu
et al. [63] considered minimum heat consumption criterion for optimisation of heat
transfer process and conducted comparison between different parameters such as
minimum heat consumption (MHC), minimum entransy dissipation (MSD) and
minimum power consumption (MPC). Figure 3.53 shows the comparison of heat
transfer characteristics and flow resistance. Thus, it is clear that MPC approach
performed well at the same power. There was not too much deviation in flow
resistance as these three lines are nearly coinciding to each other. The comparison
of transport efficiency of both smooth tube and enhanced tubes has been presented in
Fig. 3.54 with Reynolds number $Re = 200$ and $Re = 400$, respectively. They
observed and proposed MHC approach for circular tube as it has transport efficiency
far better than that of smooth tube.

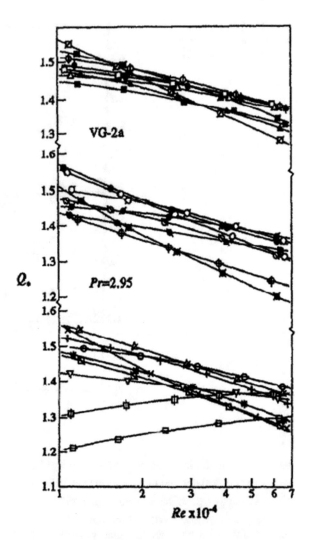

Fig. 3.35 Increased heat rate vs. Reynolds number [27]

The earlier research has modified the performance comparison of LaHaye et al. [64] for comparing the plate-finned surfaces [65], un-finned surfaces and sand-roughened surfaces. He proposed that the proposed model can work for any heat exchanger. For modified tubes, the heat transfer coefficient h and friction factor f magnitudes have been represented as h_n and f_n to the base plate area A_b. The effect of fins of any shape has been involved in h_n and f_n. Similarly, he also proposed changes in other parameters like Reynolds number, mass velocity, friction factor and hydraulic diameter. These changes were compared with Kays and London parameters and have been represented in Table 3.9. Further, the comparison between Colburn j factor and friction f and proposed Soland j_n factor and friction factor f_n has been shown in Fig. 3.55. Soland j_n and f_n have higher curve lines than the existing one. It indicates better performance. He dealt with the objective of higher

Fig. 3.36 Augmentation entropy generation number vs. Reynolds number [27]

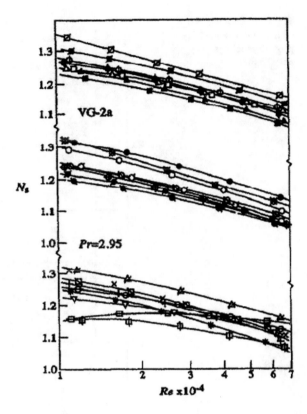

heat transfer rate by considering the monotonic interrelation between ε and NTU. At constant fluid properties and flow rate, an increment in $A.h_n$ causes increase in NTU and heat transfer rate as the relation is $\text{NTU} = \frac{Ah_n}{\omega C_P}$. On solving the equations, $\frac{\text{NTU}}{V} \propto \frac{j_n Re_n}{D_n^2}$ has been observed. Thus, modification helped in comparatively easier and simpler plotting of performance parameters $\frac{f_n Re_n^3}{D_n^4}$ and $\frac{j_n Re_n}{D_n^2}$. This has been shown in Fig. 3.56 from which performance comparison between two heat exchangers can be done. Soland (1975) considered four cases for performance evaluation.

Case 1: Volume and shape of heat exchangers are same. This case was shown in Fig. 3.56 by comparison of point "o" and point "a".

Case 2: Volume and pumping power of heat exchanger are same. This was represented as ordinate values of point o and point b, respectively. Figure 3.57 depicts the case $P = $ constant and $V = $ constant.

Case 3: Pumping power and number of transfer units are same. It means comparison of heat exchanger size. This can be represented by taking ratio of abscissa or ordinate at points o and c. This case has been represented in Fig. 3.58.

Case 4: Volume and number of transfer units are same which means that it compares pumping power for surfaces. This was represented as the ratio of abscissa of point at o and d and this is presented in Fig. 3.59.

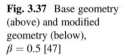
Fig. 3.37 Base geometry
(above) and modified
geometry (below),
$\beta = 0.5$ [47]

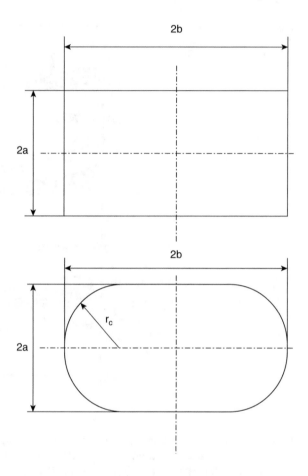

The typical performance comparison results have been shown in Fig. 3.60. In Fig. 3.61, the plate-fin surface performance parameter curves have been shown and all plate-fin surfaces were compared to smooth surface having the same nominal diameter (D_n) and without any enhancement. For plate-fin surface, Kay and London [65] presented Table 3.10. A research investigation has presented Table 3.11 to conclude the performance of surface and presented it in order of decreasing performance. He concluded that wavy-fin plate-fin 17.8–3/8w with plate spacing of 0.41 in. was best among other plate-fin heat exchangers.

Zimparov and Penchev [67] worked on the extended performance evaluation criteria to access the advantages of deeply spirally corrugated tubes over smooth tube for shell-and-tube heat exchangers. Their objective was to determine the effect of the characteristic parameters of the spirally corrugated tubes e/D_i, p/e and β_* on the thermodynamic efficiency. They investigated and plotted Figs. 3.62, 3.63, 3.64 and 3.65. They obtained their heat transfer and pressure drop data from the studies of Newson and Hodgson [66] and presented Tables 3.12, 3.13 and 3.14 of characteristic parameters of tubes extracted from Newson and Hodgson [66] literature. Zimparov

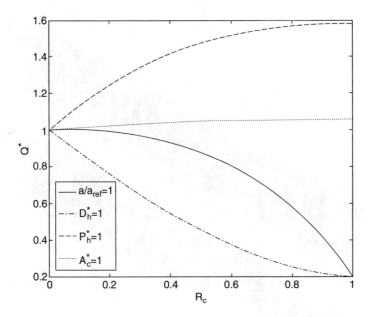

Fig. 3.38 Normalised heat duty for $\beta = 1$, $Br_{ref} = 0.1$ [47]

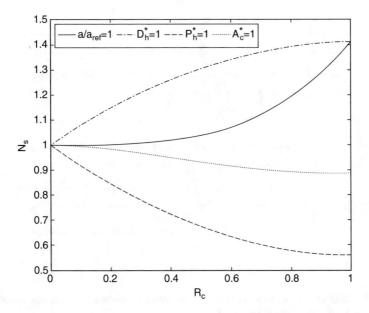

Fig. 3.39 Entropy generation number *NS* for $\beta = 1$, $Br_{ref} = 0.1$, Case FG1a [47]

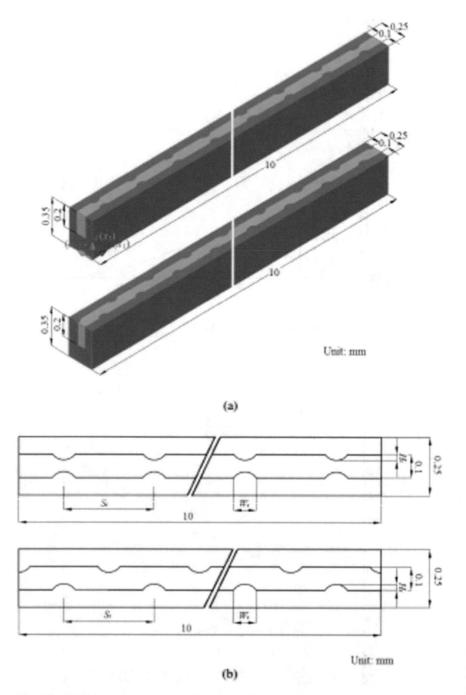

Fig. 3.40 Microchannel heat sinks with fan-shaped ribs on sidewalls. (**a**) Computational domain. (**b**) Geometric parameters of fan-shaped ribs [50]

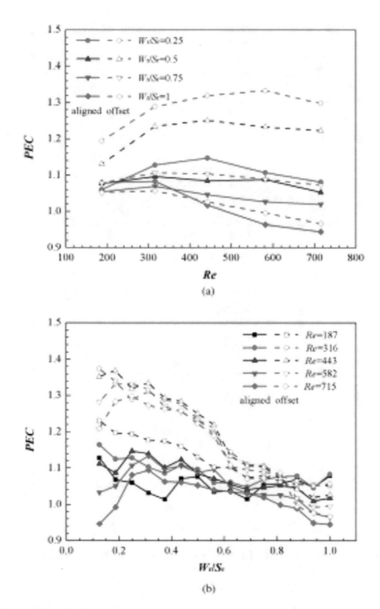

Fig. 3.41   Effects of W_r/S_r on PEC ($W_c = 0.1$ mm, $H_r = 0.025$ mm and $S_r = 0.4$ mm). (**a**) PEC vs. Re and (**b**) PEC vs. W_r/S_r [50]

and Penchev [67] concluded that corrugated tubes having small pitches and helix angle comparable to 90° have higher thermodynamic efficiency. However, higher pressure drop or pumping power was overshadowed by high heat transfer coefficient and ultimately got better thermodynamic performance. They observed that periodic disruption of boundary layer is more important than the rotational flow generated by

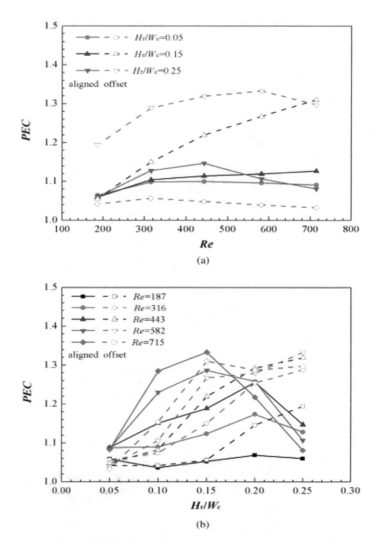

Fig. 3.42 Effects of H_r/W_c on PEC ($W_c = 0.1$ mm, $S_r = 0.4$ mm and $W_r = 0.1$ mm). (**a**) PEC vs. Re and (**b**) PEC vs. H_r/W_c [50]

swirl flow. These observations were for horizontal condensers whereas for vertical condensers higher ratio (e/D), small pitches and moderate β_* of tubes gave good satisfactory results.

Zimparov et al. [69] studied the effect of deep corrugated tubes with twisted tape insert on hydrothermal performance characteristics for a single-phase turbulent flow. The experimental work was done in the range of Reynolds number $3500 \le Re \le 50{,}000$, height-to-diameter ratio $0.053 \le e/D_i \le 0.089$ and pitch-to-

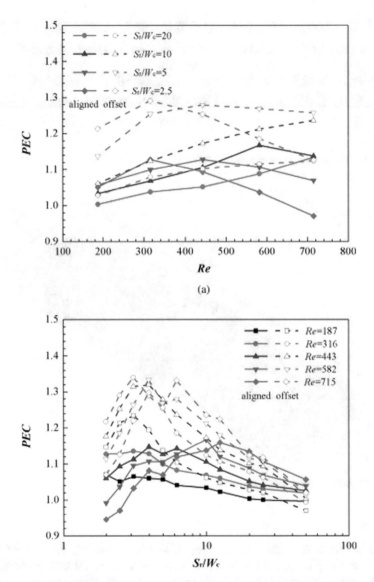

Fig. 3.43 Effects of S_t/W_c on PEC ($W_c = 0.1$ mm, $W_r = 0.1$ mm and $H_r = 0.025$ mm). (**a**) PEC vs. Re and (**b**) PEC vs. S_t/W_c [50]

height ratio $6.8 \leq p/e \leq 11.0$. They investigated the effect of geometrical parameters and Reynolds number on outside heat transfer coefficient (E_O):

$$E_o = \frac{h_{O,R}}{h_{O,S}} \tag{3.28}$$

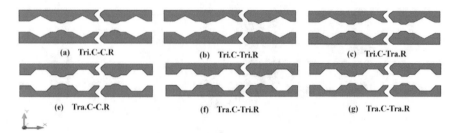

(a) Tri.C-C.R   (b) Tri.C-Tri.R   (c) Tri.C-Tra.R

(e) Tra.C-C.R   (f) Tra.C-Tri.R   (g) Tra.C-Tra.R

Fig. 3.44 Types of single microchannels [62]

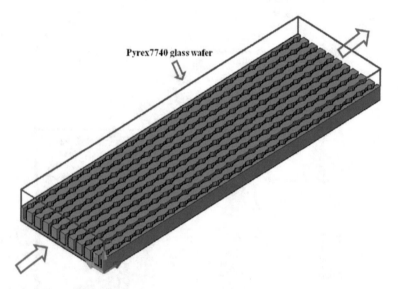

Pyrex7740 glass wafer

Fig. 3.45 Schematic diagram of micro heat sink [62]

where R stands for enhanced tube and S stands for smooth tube. Table 3.15 represents the geometrical parameters of corrugated tubes. Table 3.16 shows the uncertainties of measured physical quantities and calculated parameter. Figure 3.66 shows the variation of the friction factor with Reynolds number for the corrugated tubes alone (without twisted tape). Figure 3.67 illustrates the variation of the heat transfer coefficient for the corrugated tube alone. The experimental results revealed that the value of E_O was higher for corrugated tube used individually without twisted tape in comparison to compound enhanced technique. The value of E_O was decreased at the smaller relative pitches (H/D_i) for the combination of corrugated tube and twisted tape insert.

Yilmaz et al. [34] studied the performance of heat exchanger based on second law performance evaluation criteria. They found that first design of heat exchanger was

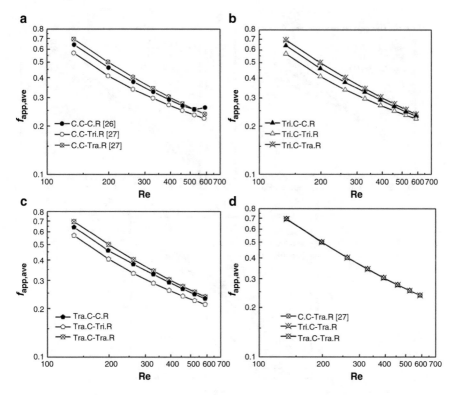

Fig. 3.46 Comparison of the apparent friction factor of the micro heat sinks with different cavities and ribs [62]

important for evaluation of PEC. Then, second law-based PEC was classified into two parts:

1. Criteria that use entropy as evaluation parameter
2. Criteria that use exergy as evaluation parameter

Table 3.17 shows the interrelation between the thermodynamic and physical parameters for performance criteria. Many investigators have given different dimensionless entropy generation rate parameters: "entropy generation number", "entropy generation unit", "non-dimensional entropy generation", "number of entropy production unit", etc. Natalini and Sciubba [70] introduced another dimensionless entropy generation number which was called Bejan number. The value of entropy generation number can vary in the range of zero to infinity. Higher values of entropy generation number mean conditions of higher losses due to heat transfer irreversibility and fluid friction irreversibility. Bejan [71], Nag and Mukherjee [72] and Bejan [15] introduced non-dimensional entropy generation number (N_S) by dividing entropy generation flow rate(mC_p). Witte and Shamsundar [73] and London and Shah [74] evaluated design of heat exchanger and costs of irreversibility in the heat

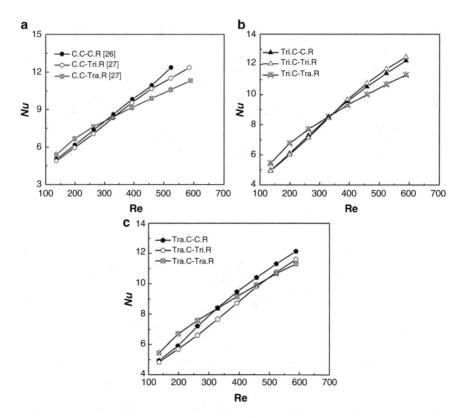

Fig. 3.47 Comparison of Nusselt number of microchannels with different cavities and ribs [62]

exchanger devices. Bruges [75] and Reistad [76] introduced dimensionless number rational (second law) effectiveness to measure the heat exchanger irreversibility and can be calculated by the following relationship:

$$\varepsilon_R = \frac{\dot{m}_C \left(e_{x,\text{out}} - e_{x,\text{in}}\right)_c}{\dot{m}_H \left(e_{x,\text{in}} - e_{x,\text{out}}\right)_H} \tag{3.29}$$

Bergles et al. [8, 77], Webb and Eckert [9], Webb and Scott [21] and Shah [2] attempted to simplify the problem related to PEC and evaluated the comparative performance of existing surfaces. He studied the algebraic equations of PEC and established the correlation between Stanton number (St) and fanning friction factor (f) of the augmented surfaces. Figure 3.68 shows the effect of reduced flow rate $\left(\frac{W}{W_S} < 1\right)$ on heat exchanger. Kays and London [65], LaHaye et al. [64], Cox and Jallouk [79] and Soland et al. [6] analysed and compared the performance of different compact heat exchangers. The V/V_S versus G_S/G for the two roughness geometries operated at $e^+ = 20$ is shown in Fig. 3.69.

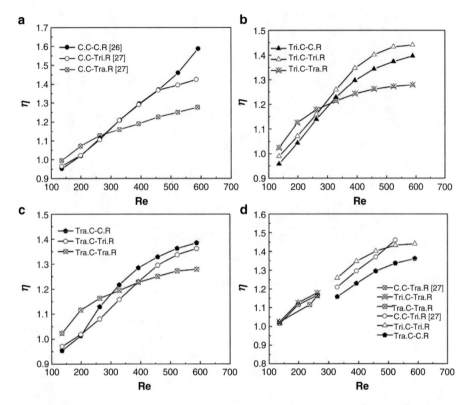

Fig. 3.48 Comparison of thermal enhanced factor of microchannels with different cavities and ribs [62]

Chakraborty and Ray [80] studied the combined first and second law-based thermal-hydraulic performance of laminar fully developed flow through square ducts with rounded corners. Objective functions were taken from suggestions of Webb and Bergles [81]. These ducts have been subjected to four specific geometric constraints and three different thermal and (or) hydraulic constraints. They observed these results: optimal configuration of ducts strongly depends on geometric parameters, thermal-hydraulic constraints and objective functions.

Figure 3.70 shows variations of objective function, R^* and N_S with R_C for FN-2 and FN-3 criteria and H1 boundary condition. They also studied the effect of pressure drop or friction factor on irreversibility distribution ratio and augmentation entropy generation number. Figure 3.71 shows variations of objective function and N_S with R_C for FG-1a criterion and H2 boundary condition. Figure 3.72 shows variations of objective function and N_S with R_C for FG-1b criterion and H2 boundary condition. The geometry of the square duct with rounded corner is shown in Fig. 3.73. Webb and Eckert [9], Bergles et al. [77], and Webb [78] have introduced several performance evaluation criteria (PEC) to assess overall improvement in thermal-hydraulic behaviour. Table 3.18 presents the various performance

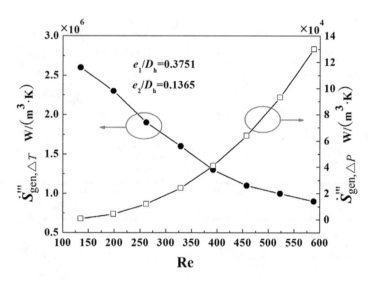

Fig. 3.49 Entropy generation rate vs. Reynolds number [62]

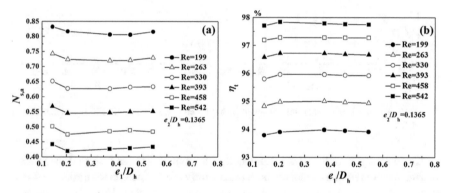

Fig. 3.50 Augmentation entropy generation (**a**) and transport efficiency of thermal energy (**b**) vs. relative cavity height [62]

evaluation criteria and objective functions, as suggested by Webb and Bergles [81], based on first law analysis.

Dong et al. [82] conducted an experiment on nine crossflow heat exchangers having novel lanced fin and flat tube. Air was used as a working fluid having Reynolds number in the range of 800–6500. By using the effectiveness-NTU method, air-side thermal performance data was analysed. They examined the effect of geometrical parameters such as fin spaces and fin lengths on heat transfer and pressure drop in terms of the Colburn j factor and Fanning friction factor f, as a function of Re.

Lin and Lee [83] carried out an experiment on pin-fin array under crossflow and evaluated the entropy generation rate based on second law analysis. They observed

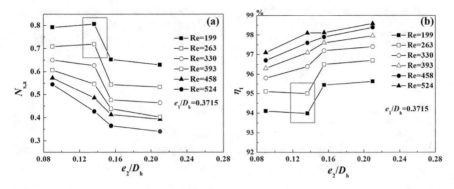

Fig. 3.51 Augmentation entropy generation number (**a**) and transport efficiency of thermal energy (**b**) vs. relative rib height [62]

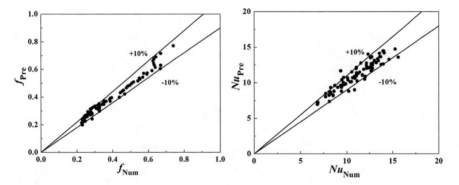

Fig. 3.52 Comparison of numerical and theoretical values, (**a**) friction factor f, (**b**) Nusselt number Nu [62]

Table 3.8 The relations among physical quantities entropy, enerty and entransy as well as entropy generation, heat consumption and entransy dissipation [63]

Variable/unit	Transport quantity	Irreversible nature	Variable differential
s J/(kg K)	$\frac{q}{T}$	$\frac{\lambda(\nabla T)^2}{T^2}$	$ds = C\frac{dT}{T}$
$e = h - T_0 s$ J/kg	q	$\frac{\lambda(\nabla T)^2}{T}$	$de = cdT - T_0 ds$
z (J K)/kg	qT	$\lambda(\nabla T)^2$	$dz = cTdT$

that heat transfer rate could be enhanced and heat transfer irreversibility could be reduced by increasing crossflow fluid velocity. They found optimal value of Reynolds number, heat exchanger design and operational conditions for minimisation of entropy generation. They compared the different thermo-hydrodynamic characteristic parameters between the in-line and the staggered fin alignment. Figure 3.74 shows the alignment of fin arrays schematically. They evaluated the effectiveness

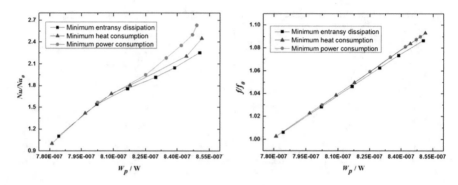

Fig. 3.53 Heat transfer process optimisation under different methods. (**a**) Heat transfer performance and (**b**) flow resistance [63]

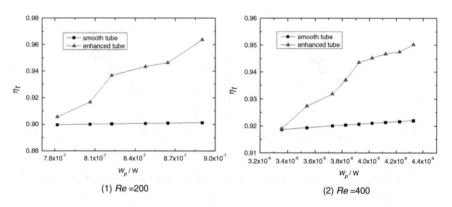

Fig. 3.54 Transport efficiency of thermal energy of smooth tube and enhanced tube under different power consumption [63]

Table 3.9 Definitions (Soland 1975)

Quantity	Kays and London [4, 14, 38, 65]	Proposed
Hydraulic diameter of radius	$r_h \equiv \frac{A_c L}{A_T}$ (1a)	$D_n \equiv \frac{4 A_y L}{A_b} = \frac{4V}{A_b}$ (1b)
Mass velocity	$G_c \equiv \frac{w}{A_c}$ (2a)	$G_n \equiv \frac{w}{A_{cF}}$ (2b)
Reynolds number	$Re \equiv \frac{4 G_c r_h}{\mu}$ (3a)	$Re_n \equiv \frac{G_n D_n}{\mu}$ (3b)
Friction factor	$f \equiv \frac{\Delta p_E}{\frac{L}{r_h} \frac{G_c^2}{2 \rho g_0}}$ (4a)	$f_n \equiv \frac{\Delta p_F}{4 \frac{L}{D_n} \frac{G_n^2}{2 \rho g_0}}$ (4b)
Heat transfer coefficient	$h \equiv \frac{q/\eta_o A_T^{*}}{\Delta T}$ (5a)	$h_n \equiv \frac{q/A_b}{\Delta T}$ (5b)
Nusselt number	$Nu \equiv \frac{4 r_h h}{k}$ (6a)	$Nu_n \equiv \frac{b_n D_n}{k}$ (6b)

Fig. 3.55 Comparison of
Colburn j factors (j and j_n)
and comparison of friction
factor (f and f_n)

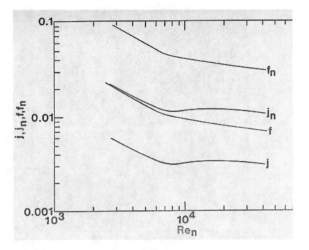

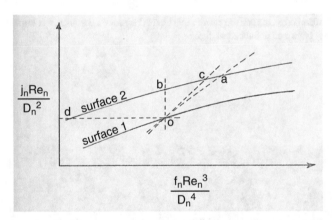

Fig. 3.56 Performance parameter curves for two surfaces showing points used in simple
comparison

of both type of fins and plotted a graph between effectiveness and Reynolds number
which is shown in Fig. 3.75:

$$\varepsilon = \left[\tfrac{\pi}{4}NVKD^2 m \tanh(ml) + hw\left(A - \tfrac{\pi}{4}NVD^2\right)\right] \Big/ {}_{hwA} \tag{3.30}$$

Table 3.19 shows the values of C and n for different ranges of Reynolds number
which helps to find the value of the Nusselt number and friction factor. Bejan [16]
and Lee and Lin [84] analysed the performance and design of fins based on second
law analysis to minimise the entropy generation rate and to maximise the available
energy of the system:

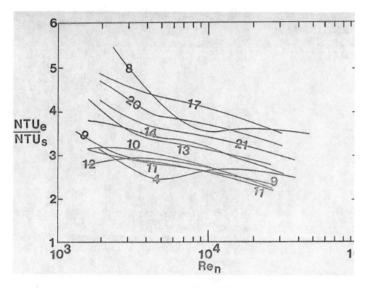

Fig. 3.57 Performance comparison results (NTUe/NTUs) for case constant volume and pumping power for both plain and enhanced surfaces

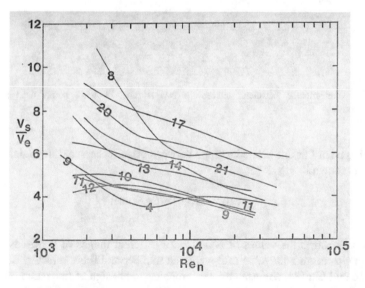

Fig. 3.58 Performance comparision results (V_s/V_e) for case NTU and pumping power constant for both enhanced and plain surfaces

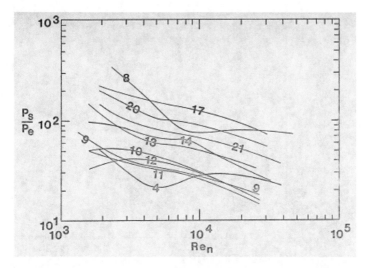

Fig. 3.59 Performance comparison results (P_s/P_e) for constant NTU and volume for both plain and enhanced tubes

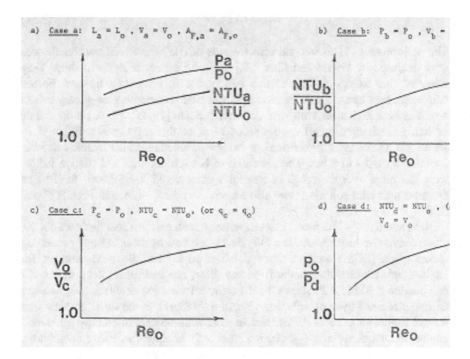

Fig. 3.60 Typical performance comparison results

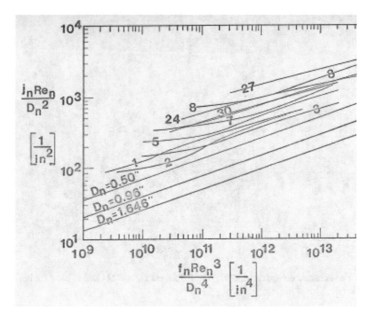

Fig. 3.61 Performance parameter curve for plate-fin surfaces

$$Nu = CRe_D^n Pr^{0.36} \qquad (3.31)$$

The performance evaluation and effectiveness of convective heat transfer devices were analysed by Prasad and Shen [32] by using exergy analysis method. They observed that exergy destructed due to increase in fluid flow friction. Forced convective heat transfer passive technique was used to minimise the exergy loss in a tubular heat exchanger with wire-coil insert. Bejan [19, 20, 85, 86] proposed an evaluation technique based on the second law of thermodynamics. Figure 3.76 shows the variation of dimensionless exergy destruction with Reynolds number for different types of tubes. The variations of Nusselt numbers and friction factors were measured over the range of Reynolds number (35,000–92,000) for 14 mm diameter tube with coil insert and also for smooth tube as shown in Figs. 3.77 and 3.78.

Ratts and Raut [87] carried out an experiment with uniform heat flux on a single-phase convective heat transfer in a fully developed flow by using entropy generation minimisation (EGM) method. They obtained an optimal Reynolds number for laminar and turbulent flow under fixed mass flow rate and fixed total heat transfer rate condition. Table 3.20 shows the different cross section of tubes, heat transfer, friction factor and hydraulic diameter. Figure 3.79 illustrates the variation of ratio of Reynolds number to optimal Reynolds number with the ratio of entropy generation number to minimum entropy generation number for laminar flow and turbulent flow.

Table 3.10 Plate-fin surfaces [65]

General surface type	Plate spacing (b) (in.)	Surface designation	Surface numbered in figures as
Plain plate-fin	.47	5.3	1
	.41	6.2	2
	.82	9.03	3
	.25	11.1	4
	.48	11.1(a)	5
	.33	14.77	6
	.42	15.08	7
	.25	19.86	8
Louvered plate-fin	.25	3/8—6.06	9
	.25	3/8(a)—6.06	10
	.25	1/2—6.06	11
	.25	1/2(a)—6.06	12
	.25	3/8—8.7	13
	.25	3/8(a)—8.7	14
	.25	3/16—11.1	15
	.25	1/4—11.1	16
	.25	1/4(b)—11.1	17
	.25	3/8—11.1	18
	.25	3/8(b)—11.1	19
	.25	1/2—11.1	20
	.25	3/4—11.1	21
	.25	3/4(b)—11.1	22
Strip-fin plate-fin	.25	1/4(s)—11.1	23
	.49	3/32—12.22	24
	.41	1/8—15.2	25
Wavy-fin plate-fin	.41	11.44—3/8W	26
	.41	17.8—3/8W	27
Pin-fin plate-fin	.24	AP—1	28
	.40	AP—2	29
	.75	PF—3	30
	.50	PF—4(F)	31
	.51	PF—9(F)	32

Convective heat transfer from a flowing fluid in a duct based on second law analysis was analysed by Nag and Kumar [31]. Duct was subjected to constant heat flux boundary condition. They observed that entropy generated was a function of the initial temperature difference and the frictional pressure drop. They optimised the value of initial temperature difference so that loss of available energy (entropy generated) could be minimised. Golem and Brzustowski [88] investigated the irreversibility of heat exchanger using Reistad effectiveness, which became equal to unity in the limiting case of a reversible heat exchanger. Mukherjee et al. [89]

Table 3.11 Plate-fin exchanger with their plate spacing and surface designation [66]

No.	D_i	e_i (mm)	p_1 (mm)	e/D_i	p/e	β	β_*
S3	30.02	2.209	15.88	0.074	7.19	56.0	0.672
S13	30.02	1.840	41.24	0.061	10.31	29.7	0.330
S17	30.02	1.816	11.24	0.060	6.19	64.5	0.717
S18	30.02	2.835	22.58	0.094	7.96	46.2	0.513
S44	30.02	1.479	15.46	0.049	10.45	56.7	0.630
S46	30.02	2.896	15.79	0.096	5.45	56.1	0.623
RM41	30.02	2.310	32.05	0.077	13.87	36.3	0.403
R42	30.02	2.366	17.67	0.079	17.67	53.1	0.590

Fig. 3.62 Variation of Q_* and N_s/Q_* vs. Re

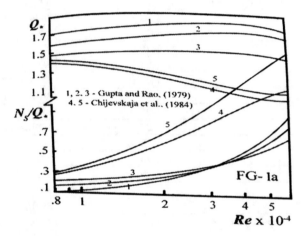

Fig. 3.63 Variation of Q_* and N_s/Q_* vs. Re [67]

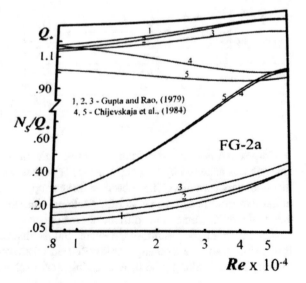

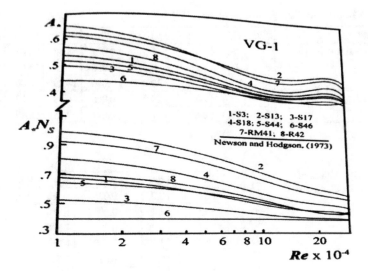

Fig. 3.64 Variation of A_* and A_*N_s vs. Re [67]

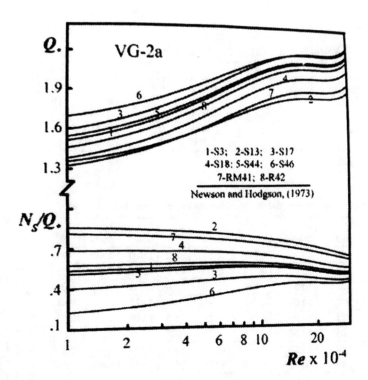

Fig. 3.65 Variation of Q_* and N_s/Q_* vs. Re [67]

Table 3.12 Characteristic parameters of tubes [67]

Authors	No.	D_i	e (mm)	p (mm)	e/D_i	p/e	β	β^*
Gupta and Rao [68]	1	23.1	1.42	6.5	0.061	4.58	84.87	0.943
	2	22.0	1.23	6.5	0.056	5.28	84.60	0.940
	3	25.0	1.40	20.0	0.056	14.29	75.69	0.841
Chijevskaja et al. (1984)	4	17.0	1.10	20.0	0.065	18.18	41.67	0.463
	5	17.0	0.80	7.0	0.047	8.75	68.49	0.761

Table 3.13 Values of constants c_f, m, c_b, n and E_o [66]

No.	c_f	m	c_b	n	E_o
S3	2.282	0.024	1.521	0	2.55
S13	1.179	0.009	1.130	0	2.19
S17	3.205	0.028	1.826	0	2.62
S18	1.872	0.020	1.217	0	2.44
S44	2.179	0.015	1.565	0	2.56
S46	2.769	0.028	2.174	0	2.56
RM41	1.436	0.013	1.087	0	2.30
R42	2.154	0.024	1.391	0	2.52

Table 3.14 Values of constants c_f, m, c_b, n and E_o [67]

Author	No.	c_f	m	c_b	n	E_o
Gupta and Rao [68]	1	0.570	0.217	0.630	0.157	1.30
	2	0.506	0.208	0.478	0.168	1.30
	3	1.063	0.101	0.799	0.101	1.10
Chijevskaja et al. (1984)	4	2.000	0	24.119	−0.286	1.10
	5	3.714	0	24.698	−0.289	1.30

Table 3.15 Geometrical parameters of the corrugated tubes [69]

No.	D_o (mm)	D_i (mm)	e (mm)	p (mm)	β (°)	t (mm)	s (mm)	e/D_i	p/e	β_*
5010	15.44	13.68	0.797	7.57	80.0	1.910	0.368	0.058	9.50	0.889
5020	15.36	13.64	0.772	6.26	81.7	1.936	0.385	0.056	8.11	0.908
5030	15.31	13.51	0.767	5.19	83.0	1.836	0.371	0.057	6.77	0.922
5040	15.73	14.04	0.797	8.73	78.8	2.021	0.360	0.057	10.95	0.875
5050	15.71	13.99	0.744	6.55	81.5	1.719	0.340	0.053	8.80	0.905
5060	15.65	14.00	0.578	4.78	83.8	1.392	0.257	0.041	8.27	0.931
6010	15.80	14.12	1.256	11.71	75.2	2.476	0.553	0.089	9.32	0.836
6020	15.73	14.02	1.122	10.56	76.5	2.434	0.499	0.080	9.41	0.850
6030	15.72	13.98	1.073	9.194	78.2	2.338	0.480	0.077	8.57	0.869

Table 3.16 Uncertainties of measured quantities and calculated parameters [69]

Parameter	Uncertainty
Water and steam temperature, T_i, T_o, T_s	±0.1%
Mean tube wall temperature, T_w	±0.5%
Pressure drop, ΔP	±5.0%
Mass flow rate of water, $\dot{m}$	±2.0%
Fanning friction factor, f	±6.5%
Condensing heat transfer coefficient, h_o	±2.5%
Inside heat transfer coefficient, h_i	±(15–20)%

Fig. 3.66 The variation of friction factor vs. Reynolds number for all corrugated tubes [69]

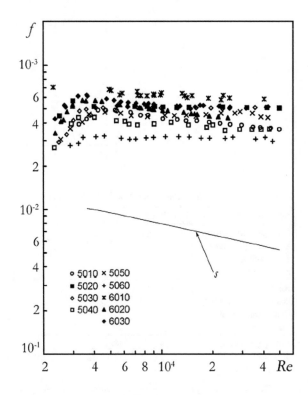

presented a second law analysis for convective heat transfer in swirling flow. Figure 3.80 presents the variation of optimum value of temperature difference against non-dimensional duty parameter for air with $A = 0.01$ and n as a parameter. They also tried to find an optimum fluid velocity corresponding to which loss of available power was minimised. Figure 3.80 shows the variation of non-dimensional

Fig. 3.67 The variation of Nusselt number vs. Reynolds number for all corrugated tubes [69]

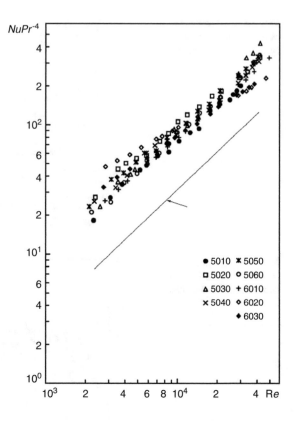

Table 3.17 The interrelation of some performance criteria [14]

Performance criteria	Relation	References
Entropy generation number and Witte-Shamsundar efficiency	$\eta W - S = 1 - \frac{T_0 c_{P,H}}{(h_{in} - h_{out})_H} N_S$	Witte and Shamsundar [73] Bejan [15–17, 19, 20, 71, 85, 86]
Rational (second law) efficiency and Witte-Shamsundar efficiency	$\frac{1 - \eta W - S}{1 - \varepsilon_R} = \left(\frac{e_{x,in} - e_{x,out}}{h_{in} - h_{out}} \right)_H$	Witte and Shamsundar [73] Bruges [75] Reistad [76] Bejan [15–17, 19, 20, 71, 85, 86]
Entropy generation number and rational (second law) efficiency	$\varepsilon_R = 1 - \frac{T_0 \dot{S}_{gen}}{\dot{m}_H (e_{x,in} - e_{x,out})_H} = 1 - \frac{T_0 c_{P,H}}{(e_{x,in} - e_{x,out})_H} N_S$	Bruges [75] Reistad [76] Bejan [15–17, 19, 20, 71, 85, 86]
Entropy generation number and merit function	$M = \frac{\overline{Nu}\left(1 - \frac{\theta_h}{\theta_w}\right)(1 - \sigma_a)}{\overline{Nu}\left(1 - \frac{\theta_h}{\theta_w}\right)(1 - \sigma_a) + \frac{1}{\pi}\left(\frac{D}{L}\right)\frac{\sigma_a}{\sigma} N_S}$	Mukherjee et al. [89]
Entropy generation number and quality of energy transformation (HERN)	$Y_s = 1 - \frac{N_s}{N_{s,max}}$	Sekulic [44]

Fig. 3.68 Effect of reduced flow rate on Q/Q_S for $UA/U_SA_S = 1$ [78]

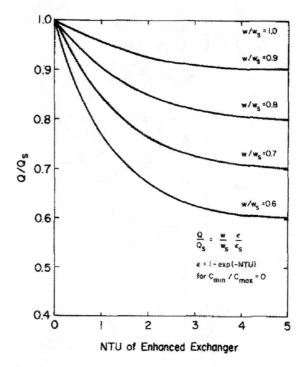

NTU of Enhanced Exchanger

entropy generation number (N_S) with non-dimensional temperature difference (τ) for $n = 60, A = 0.01$ and J as a parameter. Figure 3.81 illustrates the variation of N_S with τ for $J = 2 \times 10^{-4}, A = 0.01$ and n as a parameter where J is the non-dimensional duty parameter defined as

$$J = \frac{f^{1/2}Q}{\rho a \left(C_p T_0\right)^{3/2}} \tag{3.32}$$

Sahin [90] studied the entropy generation rate and pumping power for improvement of heat transfer enhancement in a turbulent fluid flow through a smooth pipe subjected to constant heat flux. He determined the optimal value of heat transfer coefficient and friction factor by considering the viscosity as an important parameter as a function of temperature. It was observed that viscosity variation has affected both the entropy generation and pumping power. Different thermophysical properties of water and glycerol have been shown in Table 3.21. Figure 3.82 shows the optimization of convective heat transfer. Variation of dimensionless entropy generation with modified Stanton number for water and glycerol (Fig. 3.83). Additional references in ths regard are [91–95].

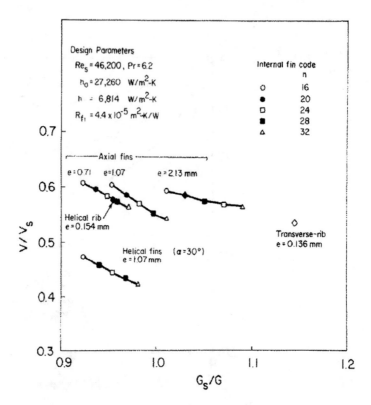

Fig. 3.69 V/V_S vs. G_S/G for several water-side enhancement, $P/P_S = Q/Q_S = 1$, with $w/w_s = 1$ [78]

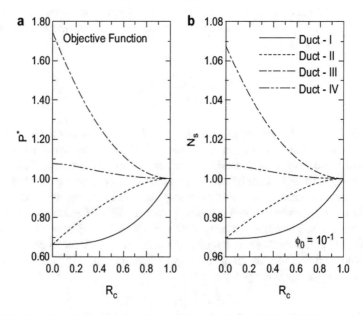

Fig. 3.70 Variations of objective function and N_S with R_c for FN-2 and FN-3 criteria and H1 boundary condition. (a) P^*, (b) N_S for $\emptyset_o = 0.1$ [81]

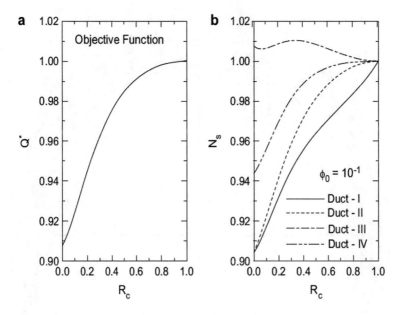

Fig. 3.71 Variation objective function and N_s with R_c for FG-1a criterion and H2 boundary condition. (**a**) Q^*, (**b**) N_s for $\emptyset_o = 0.1$ [81]

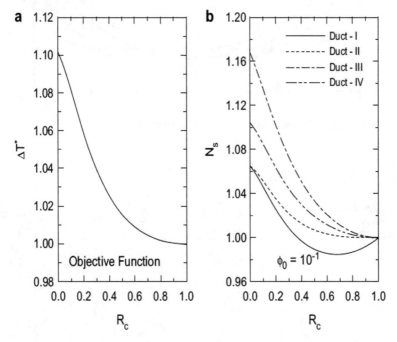

Fig. 3.72 Variation of objective function and N_S with R_C for FG-1b criterion and H2 boundary condition. (**a**) ΔT, (**b**) N_S for $\emptyset_o = 0.1$ [81]

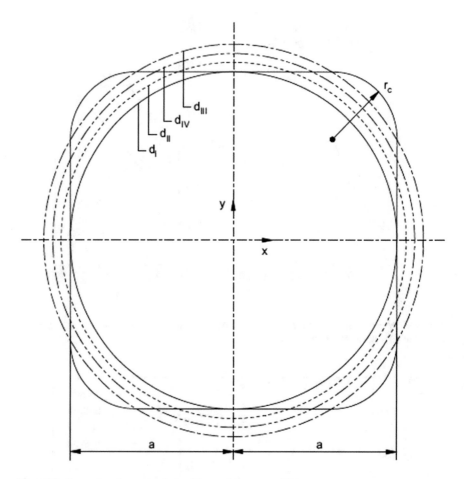

Fig. 3.73 Geometry of a square duct with rounded corners [80]

Table 3.18 List of fixed parameters and objective functions [80]

Criterion	L^*	$\dot{m}^*$	P^*	$\dot{Q}^*$	ΔT^*	Objective
FG-1a	1	1			1	$\uparrow \dot{Q}^*$
FG-1b	1	1		1		$\downarrow \Delta T^*$
FG-2a	1		1		1	$\uparrow \dot{Q}^*$
FG-2b	1		1	1		$\downarrow \Delta T^*$
FN-1		1	1	1		$\downarrow L^*$
FN-2	1			1	1	$\downarrow L^*$
FN-3	1			1	1	$\downarrow P^*$
VG-2a		1	1		1	$\downarrow L^*$
VG-2b		1	1	1		$\downarrow \Delta T^*$

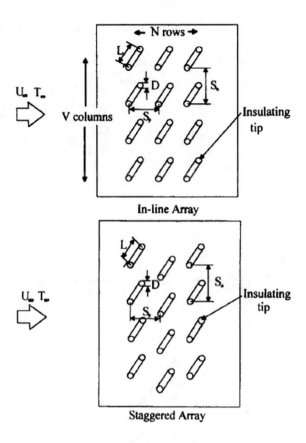

Fig. 3.74 Schematic drawing of the pin-fin arrays (top) in-line alignment (bottom) staggered alignment [83]

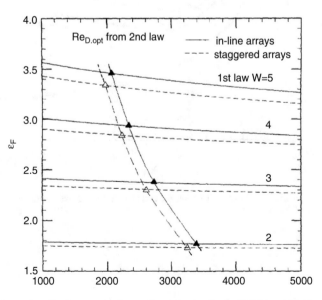

Fig. 3.75 Shows fin effectiveness vs. Re_D under various values of W. $M = 100$, $B = 10^{-13}$, $S_p/S_n = 1$, $S_n/D = 1.25$, $N = 20$, $V = 10$ [83]

Table 3.19 Values of C and n used to find Nusselt number values [83]

Re_D range	Staggered arrays			In-line arrays	
		C	n	C	n
10–100		0.8	0.4	0.9	0.4
100–1000		0.71	0.5	0.52	0.5
1000–2×10^5	$(S_n/S_p) < 2$	$0.35(S_p/S_n)^{-0.2}$	0.6	0.27	0.63
	$(S_n/S_p) > 2$	0.4	0.6		
2×10^5–10^6		$0.031(S_p/S_n)^{-0.2}$	0.88	0.03	0.8

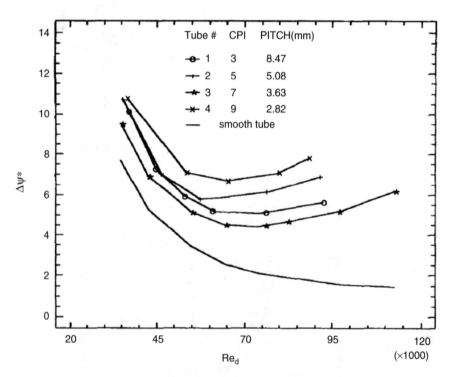

Fig. 3.76 Dimensionless exergy destruction for coiled-wire inserts (wire diameter = 0.8 13 mm)
[32]

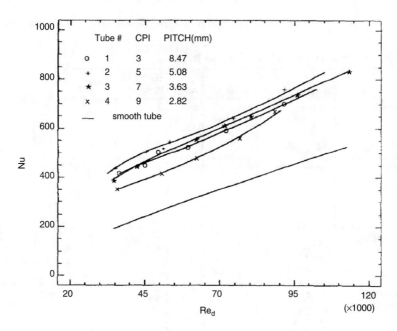

Fig. 3.77 Nusselt number with coiled-wire inserts (wire diameter = 0.813 mm) [32]

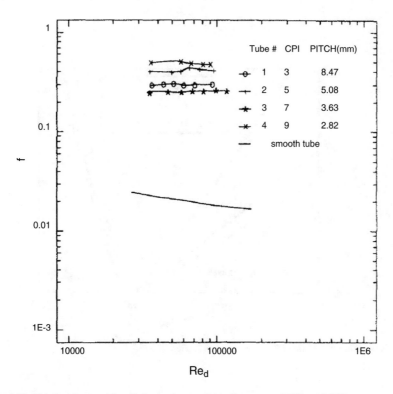

Fig. 3.78 Friction factor with coiled-wire inserts (wire diameter = 0.813 mm) [32]

Table 3.20 Different cross section of the tubes [87]

Cross section	Diagram	Perimeter	Area	Hydraulic diameter	$Nu^{\dagger}$	$fRe^{\dagger}$
Circular		πa	$\frac{\pi a^2}{4}$	a	4.36	64
Square		$4a$	a^2	a	3.61	57
Rectangle		$2a\left(1+\frac{b}{a}\right)$	$a^2\left(\frac{b}{a}\right)$	$\frac{2a}{\left(\frac{a}{b}+1\right)}$	$b/a = 2$: 4.12 $b/a = 8$: 6.49	$b/a = 2$: 62 $b/a = 8$: 82
Equilateral triangle		$3a$	$\frac{\sqrt{3}}{4}a^2$	$\frac{\sqrt{3}}{3}a$	3.11	53

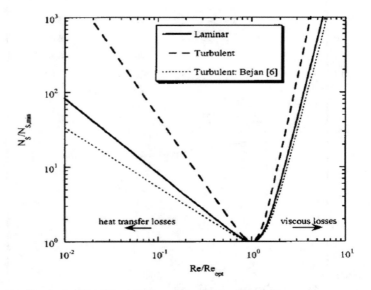

Fig. 3.79 Entropy generation for circular cross section [87]

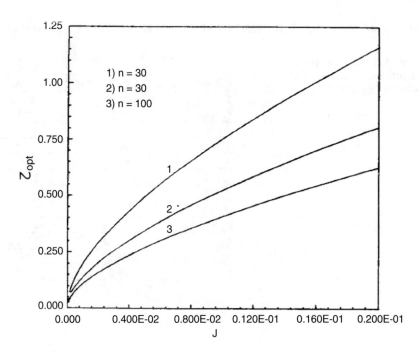

Fig. 3.80 Variation of τ_{opt} with J having n as a parameter [31]

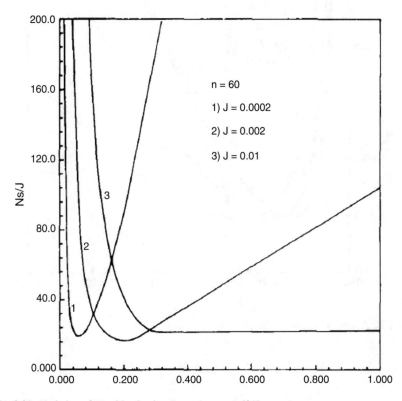

Fig. 3.81 Variation of N_s with τ having J as a parameter [31]

Table 3.21 Thermophysical properties and parameters for water and glycerol [90]

	Water	Glycerol
b (N·s·m^{-2}·K^{-1})	8.9438×10^{-6}	0.0185
B	4700	23,100
C_p (J·kg^{-1}·K^{-1})	4182	2438
D (m)	0.1	1
k (W·m^{-1}·K^{-1})	0.6	0.264
n	8.9	52.4
T_{ref} (K)	293	293
T_0 (K)	293	293
$\bar{U}$ (m·s^{-1})	0.02	2
μ_{ref} (N·s·m^{-2})	9.93×10^{-4}	1.48
Π_1	0.0–1.0	0.0–1.0
ρ (kg·m^{-3})	998.2	1260
τ	0.0–0.1	0.0–0.1

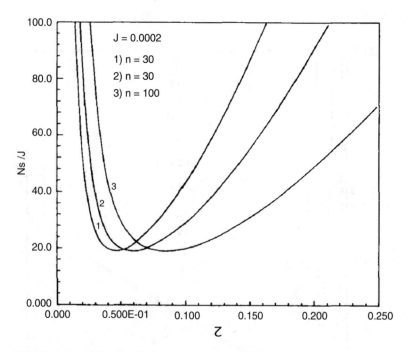

Fig. 3.82 Variation of N_s with τ having n as a parameter [31]

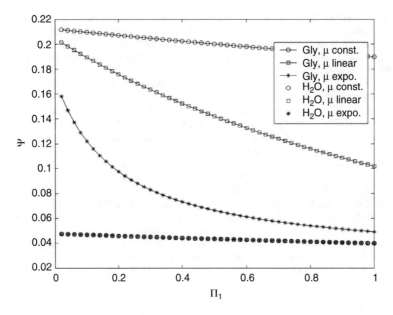

Fig. 3.83 Dimensionless entropy generation ψ versus modified Stanton number Π_1 for water and glycerol with three cases of viscosity dependence ($\tau = 0.05$) [90]

References

1. Webb RL, Haman LL, Hui TS (1984) Enhanced tubes in electric utility steam condensers. In: Sengupta S, Mussalli YG (eds) Heat transfer in heat rejection systems. ASME Symp 37:17–26
2. Shah RK (1978) Compact heat exchanger surface selection methods. Heat Transfer:193–199
3. Cowell TA (1990) A general method for the comparison of compact heat transfer surfaces. J Heat Transfer 112(2):288–294
4. Kays WM, London AL (1984) Compact heat exchangers, McGraw-Hill.
5. Soland JG, Rohsenow W, Mack W (1976) Performance ranking of plate-fin heat-exchanger surfaces. Mech Eng 99(6):110, ASME, New York
6. Soland JG, Mack WM, Rohsenow WM (1978) Performance ranking of plate-fin heat exchanger surfaces. J Heat Transfer 100(3):514–519
7. Yilmaz M, Comakli O, Yapici S, Sara ON (2005) Performance evaluation criteria for heat exchangers based on first law analysis. J Enhanc Heat Transf 12(2):121–158
8. Bergles AE, Bllumenkrantz AR, Taborek (1973) Performance evaluation criteria for enhanced heat transfer surfaces. Heat Transfer V-II 1974. Jpn Soc Mech Eng 2:234–238
9. Webb RL, Eckert ER (1972) Application of rough surfaces to heat exchanger design. Int J Heat Mass Transfer 15(9):1647–1658
10. Shah RK, London AL (1978) Laminar flow forced convection in ducts, advances in heat transfer, BH, Elsevier.
11. Sundén B (1999) Heat transfer and fluid flow in rib-roughened rectangular ducts. In: Heat transfer enhancement of heat exchangers. Springer, Dordrecht
12. Shah RK (1983) Compact heat exchanger surface selection, optimization, and computer aided thermal design. In: Low Reynolds number flow heat exchangers. Hemisphere, Washington, DC, pp 845–876

13. Kakaç S (1998) Introduction to heat transfer enhancement PEC based on first law of thermo-dynamics. In: Kakaç S (ed) Energy conservation through heat transfer enhancement of heat exchangers. Nato Advanced Study Institute 12, Num 2005 155
14. Kays WM, London AL (1950) Heat transfer and flow friction characteristics of some compact heat exchanger surfaces. Trans ASME 72:1075–1097
15. Bejan A (1977) The concept of irreversibility in heat exchanger design: counterflow heat exchangers for gas-to-gas applications. J Heat Transfer 99(3):374–380
16. Bejan A (1982) Entropy generation through heat and fluid flow. Wiley, New York
17. Bejan A (1996) Entropy generation minimization. CRC Press, Boca Raton, FL
18. Bejan A, Pfister PA Jr (1980) Evaluation of heat transfer augmentation techniques based on their impact on entropy generation. Lett Heat Mass Transfer 7(2):97–106
19. Bejan A (1980a) Discussion: "a parametric analysis of the performance of internally finned tubes for heat exchanger application" (Webb, RL, and Scott, MJ, 1980, ASME J. Heat Transfer, 102, pp. 38–43). J Heat Transfer 102:586–587
20. Bejan A (1980b) Second law analysis in heat transfer, Energy 5(8–9):720–732
21. Webb RL, Scott MJ (1980) A parametric analysis of the performance of internally finned tubes for heat exchanger application. J Heat Transfer 102(1):38–43
22. Ouellette WR, Bejan A (1980) Conservation of available work (exergy) by using promoters of swirl flow in forced convection heat transfer. Energy 5(7):587–596
23. Nikuradse J (1950) Laws of flow in rough pipes. National Advisory Committee for Aeronautics, Washington
24. Dipprey DF, Sabersky RH (1963) Heat and momentum transfer in smooth and rough tubes at various Prandtl numbers. Int J Heat Mass Transfer 6(5):329–353
25. Kays WM, Perkins HC (1973) In: Rohsenow WM, Hartnett JP (eds) Handbook of heat transfer. McGraw-Hill, New York, Chapter 7-4
26. Chen BH, Huang WH (1988) Performance evaluation criteria for enhanced heat transfer surfaces. Int Commun Heat Mass Transfer 15:55–72
27. Zimparov VD, Vulchanov NL (1994) Performance evaluation criteria for enhanced heat transfer surfaces. Int J Heat Mass Transfer 37(12):1807–1816
28. Zimparov V (2000) Extended performance evaluation criteria for enhanced heat transfer surfaces: heat transfer through ducts with constant wall temperature. Int J Heat Mass Transfer 43(17):3137–3155
29. Zimparov V (2001a) Enhancement of heat transfer by a combination of three-start spirally corrugated tubes with a twisted tape. Int J Heat Mass Transfer 44(3):551–574
30. Zimparov V (2001b) Extended performance evaluation criteria for enhanced heat transfer surfaces: heat transfer through ducts with constant heat flux. Int J Heat Mass Transfer 44 (1):169–180
31. Nag PK, Kumar N (1989) Second law optimization of convective heat transfer through a duct with constant heat flux. Int J Energy Res 13(5):537–543
32. Prasad RC, Shen J (1993) Performance evaluation of convective heat transfer enhancement devices using exergy analysis. Int J Heat Mass Transfer 36(17):4193–4197
33. Zimparov VD, Bonev PJ, Petkov VM (2016) Benefits from the use of enhanced heat transfer surfaces in heat exchanger design: a critical review of performance evaluation. J Enhanc Heat Transf 23(5):371–391
34. Yilmaz M, Sara ON, Karsli S (2001) Performance evaluation criteria for heat exchangers based on second law analysis. Exergy 1(4):278–294
35. Zimparov V, Penchev PJ, Bergles AE (2006) Performance characteristics of some "rough surfaces" with tube inserts for single-phase flow. J Enhanc Heat Transf 13(2):117–137
36. Bishara F, Jog MA, Manglik RM (2013) Heat transfer enhancement due to swirl effects in oval tubes twisted about their longitudinal axis. J Enhanc Heat Transf 20(4):289–304
37. Shah RK, Sekulic DP (2003) Fundamentals of heat exchanger design. Wiley, New York
38. Kays WM, London AL (1998) Compact heat exchangers, 3rd edn. Krieger Publishing Company, Malabar, FL

39. Ahamed JU, Wazed MA, Ahmed S, Nukman Y, Ya TT, Sarkar MAR (2011) Enhancement and prediction of heat transfer rate in turbulent flow through tube with perforated twisted tape inserts: a new correlation. J Heat Transfer 133(4):041903
40. Lei YG, He YL, Tian LT, Chu P, Tao WQ (2010) Hydrodynamics and heat transfer characteristics of a novel heat exchanger with delta-winglet vortex generators. Chem Eng Sci 65 (5):1551–1562
41. Fan A, Deng J, Guo J, Liu W (2011) A numerical study on thermo-hydraulic characteristics of turbulent flow in a circular tube fitted with conical strip inserts. Appl Therm Eng 31 (14–15):2819–2828
42. Guo J, Fan A, Zhang X, Liu W (2011) A numerical study on heat transfer and friction factor characteristics of laminar flow in a circular tube fitted with center-cleared twisted tape. Int J Therm Sci 50(7):1263–1270
43. Fan AW, Deng JJ, Nakayama A, Liu W (2012) Parametric study on turbulent heat transfer and flow characteristics in a circular tube fitted with louvered strip inserts. Int J Heat Mass Transfer 55(19–20):5205–5213
44. Zhang X, Liu Z, Liu W (2012) Numerical studies on heat transfer and flow characteristics for laminar flow in a tube with multiple regularly spaced twisted tapes. Int J Therm Sci 58:157–167
45. Xu JL, Gan YH, Zhang DC, Li XH (2005) Microscale heat transfer enhancement using thermal boundary layer redeveloping concept. Int J Heat Mass Transfer 48(9):1662–1674
46. Esmaeilnejad A, Aminfar H, Neistanak MS (2014) Numerical investigation of forced convection heat transfer through microchannels with non-Newtonian nanofluids. Int J Therm Sci 75:76–86
47. Lorenzini M, Suzzi N (2016) The influence of geometry on the thermal performance of microchannels in laminar flow with viscous dissipation. Heat Transfer Eng 37 (13–14):1096–1104
48. Lorenzini M (2013) The influence of viscous dissipation on thermal performance of microchannels with rounded corners. La Houille Blanche 4:64–71
49. Lorenzini M, Morini GL (2011) Single-phase laminar forced convection in microchannels with rounded corners. Heat Transfer Eng 32(13–14):1108–1116
50. Chai L, Xia GD, Wang HS (2016) Parametric study on thermal and hydraulic characteristics of laminar flow in microchannel heat sink with fan-shaped ribs on sidewalls—Part 3: Performance evaluation. Int J Heat Mass Transfer 97:1091–1101
51. Dai B, Li M, Dang C, Ma Y, Chen Q (2014) Investigation on convective heat transfer characteristics of single phase liquid flow in multi-port micro-channel tubes. Int J Heat Mass Transfer 70:114–118
52. Wang Y, Houshmand F, Elcock D, Peles Y (2013a) Convective heat transfer and mixing enhancement in a microchannel with a pillar. Int J Heat Mass Transfer 62:553–561
53. Wang X, Yu J, Ma M (2013b) Optimization of heat sink configuration for thermoelectric cooling system based on entropy generation analysis. Int J Heat Mass Transfer 63:361–365
54. Escandón J, Bautista O, Méndez F (2013) Entropy generation in purely electroosmotic flows of non-Newtonian fluids in a microchannel. Energy 55:486–496
55. Khan WA, Culham JR, Yovanovich MM (2008) Optimization of pin-fin heat sinks in bypass flow using entropy generation minimization method. J Electron Packag 130(3):031010
56. Koşar A (2011) Exergo-economic analysis of micro pin fin heat sinks. Int J Energy Res 35 (11):1004–1013
57. Xia G, Zhai Y, Cui Z (2013a) Characteristics of entropy generation and heat transfer in a microchannel with fan-shaped reentrant cavities and internal ribs. Sci China Technol Sci 56 (7):1629–1635
58. Xia GD, Zhai YL, Cui ZZ (2013b) Numerical investigation of flow and heat transfer in a microchannel with fan-shaped reentrant cavities and internal ribs. Appl Therm Eng 58:52–60
59. Sharma CS, Tiwari MK, Michel B, Poulikakos D (2013) Thermofluidics and energetics of a manifold microchannel heat sink for electronics with recovered hot water as working fluid. Int J Heat Mass Transfer 58(1–2):135–151

60. Kandlikar S, Garimella S, Li D, Colin S, King MR (2005) Heat transfer and fluid flow in minichannels and microchannels. Elsevier, Amsterdam
61. Shalchi-Tabrizi A, Seyf HR (2012) Analysis of entropy generation and convective heat transfer of Al_2O_3 nanofluid flow in a tangential micro heat sink. Int J Heat Mass Transfer 55 (15–16):4366–4375
62. Zhai YL, Xia GD, Liu XF, Li YF (2015) Exergy analysis and performance evaluation of flow and heat transfer in different micro heat sinks with complex structure. Int J Heat Mass Transfer 84:293–303
63. Liu W, Jia H, Liu ZC, Fang HS, Yang K (2013) The approach of minimum heat consumption and its applications in convective heat transfer optimization. Int J Heat Mass Transfer 57 (1):389–396
64. LaHaye PG, Neugebauer FJ, Sakhuja RK (1974) A generalized prediction of heat transfer surfaces. J Heat Transfer 96(4):511–517
65. Kays WM, London AL (1964) Compact heat exchangers. McGraw-Hill, New York
66. Newson IH, Hodgson TK (1973) Scale of enhanced heat exchanger tubes. Heat Transfer 89 (3):85–89
67. Zimparov VD, Penchev PJ (2004) Performance evaluation of deep spirally corrugated tubes for shell-and-tube heat exchangers. J Enhanc Heat Transf 11(4):423–434
68. Gupta AK, Rao KM (1979) Simultaneous detection of Salmonella typhi antigen and antibody in serum by counter-immunoelectrophoresis for an early and rapid diagnosis of typhoid fever. J Immunol Methods 30(4):349–353
69. Zimparov V, Petkov VM, Bergles AE (2012) Performance characteristics of deep corrugated tubes with twisted-tape inserts. J Enhanc Heat Transf 19(1):1–11
70. Natalini G, Sciubba E (1999) Minimization of the local rates of entropy production in the design of air-cooled gas turbine blades. J Eng Gas Turbines Power 121(3):466–475
71. Bejan A (1988) Advanced engineering thermodynamics. Wiley & Sons, New York
72. Nag PK, Mukherjee P (1987) Thermodynamic optimization of convective heat transfer through a duct with constant wall temperature. Int J Heat Mass Transfer 30(2):401–405
73. Witte LC, Shamsundar N (1983) A thermodynamic efficiency concept for heat exchange devices. J Eng Power 105(1):199–203
74. London AL, Shah RK (1983) Costs of irreversibilities in heat exchanger design. Heat Transfer Eng 4(2):59–73
75. Bruges EA (1959) Available energy and the second law analysis. Butterworths, London
76. Reistad GM (1970) Availability: concepts and applications. Ph.D. thesis, University of Wisconsin, Madison
77. Bergles AE, Bunn RL, Junkhan GH (1974) Extended performance evaluation criteria for enhanced heat transfer surfaces. Lett Heat Mass Transfer 1:113–120
78. Webb RL (1981) Performance evaluation criteria for use of enhanced heat transfer surfaces in heat exchanger design. Int J Heat Mass Transfer 24(4):715–726
79. Cox B, Jallouk PA (1973) Methods for evaluating the performances of compact heat transfer surfaces. J Heat Transfer 95(4):464–473
80. Chakraborty S, Ray S (2011) Performance optimisation of laminar fully developed flow through square ducts with rounded corners. Int J Therm Sci 50(12):2522–2535
81. Webb RL, Bergles AE (1983) Performance evaluation criteria for selection of heat transfer surface geometries used in low Reynolds number heat exchangers. Iowa State University College of Engineering, Ames, IA
82. Dong J, Chen J, Chen Z, Zhou Y (2008) Air side performance of a novel lanced fin and flat tube heat exchanger. J Enhanc Heat Transf 15(1):17–30
83. Lin WW, Lee DJ (1997) Second-law analysis on a pin-fin array under crossflow. Int J Heat Mass Transfer 40(8):1937–1945
84. Lee DJ, Lin WW (1995) Second-law analysis on a fractal-like fin under crossflow. AICHE J 41:2314–2317

85. Bejan A (1978) General criterion for rating heat-exchanger performance. Int J Heat Mass Transfer 21(5):655–658
86. Bejan A (1984) Solutions manual for entropy generation through heat and fluid flow
87. Ratts EB, Raut AG (2004) Entropy generation minimization of fully developed internal flow with constant heat flux. J Heat Transfer 126(4):656–659
88. Golem PJ, Brzustowski TA (1977) Second law analysis of thermal processes. Trans CSME 4:219
89. Mukherjee P, Biswas G, Nag PK (1987) Second-law analysis of heat transfer in swirling flow through a cylindrical duct. J Heat Transfer 109(2):308–313
90. Sahin AZ (2002) Entropy generation and pumping power in a turbulent fluid flow through a smooth pipe subjected to constant heat flux. Int J Exergy 2(4):314–321
91. Kakac S, Liu H, Pramuanjaroenkij A (1997) Heat exchangers selection rating, and thermal design. CRC Press, New York
92. McClintock FA (1951) The design of heat exchangers for minimum irreversibility. ASME paper 51
93. Ouellette WR (1979) Entropy generation criterion applied to various heat transfer augmentation techniques. Doctoral dissertation, University of Colorado at Boulder
94. Perez-Blanco H (1984) Irreversibility in heat transfer enhancement. In: second law aspects thermal design ASME, HTD 33:19–26
95. Shah RK, London AL (2014) Laminar flow forced convection in ducts: a source book for compact heat exchanger analytical data. Academic Press, New York

Chapter 4
PEC for Two-Phase Flow

The PEC analysis for two-phase flow [1, 2] has been discussed in this section. Pressure drop of a two-phase fluid may reduce the mean temperature difference for heat exchange. So, PEC for single-phase flow needs to be modified in order to obtain PEC for two-phase flow. The two-phase thermal systems may be work-producing systems, work-consuming systems and heat-actuated systems.

In two-phase flow, pressure drop causes a reduction of the saturation temperature and this causes reduced LMTD (Fig. 4.1). The PEC for single-phase flow is suitably modified to account for the effect of two-phase pressure drop on the LMTD. $\frac{dT}{dp}$ at the operating pressure influences the given pressure drop on the decrease of saturation temperature, ΔT_{sat}. The magnitude of $\frac{dT}{dp}$ may be obtained from the Clausius-Clapeyron equation. $\frac{dT}{dp}$ increases as $\frac{P}{P_{cr}}$ decreases (Fig. 4.2). Refrigerant evaporates and power cycle condensers face LMTD reduction due to two-phase pressure drop. Performance improvements in the work-consuming systems may be affected by the use of enhanced surfaces in the following ways:

(a) Reduced heat transfer surface area with fixed condenser pressure
(b) Increased evaporator heat duty for fixed compressor lift, i.e. pressure difference between condenser and evaporator
(c) Reduced compressor power for fixed evaporator heat duty

The LMTD of the evaporator and condenser may be reduced (Fig. 4.3). The suction pressure to the compressor will be increased and the inlet pressure to the condenser will be reduced.

Performance improvements in the work-producing systems may be affected by the use of enhanced surfaces in the following ways:

(a) Reduced boiler and/or condenser surface area for constant turbine output.
(b) Increased turbine output for fixed boiler heat input, or fixed condenser heat rejection: This increases the boiler pressure or reduces the condensing temperature.

S. K. Saha et al., *Performance Evaluation Criteria in Heat Transfer Enhancement*, SpringerBriefs in Applied Sciences and Technology, https://doi.org/10.1007/978-3-030-20758-8_4

Fig. 4.1 Illustration of the
effect of two-phase pressure
drop on the LMTD [1, 2]

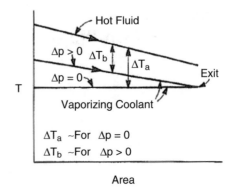

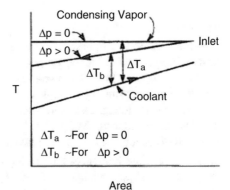

In heat-actuated systems, there may be a pump to transport liquids, but there are no compressors or turbines in the process operations. Performance improvements may be affected by use of enhanced surfaces in the following ways:

(a) Reduced heat transfer surface area for fixed operating temperatures.
(b) Increased heat exchange capacity for fixed amount of heat exchange surface area.
(c) Reduced LMTD for fixed amount of heat exchange surface area and this increases the thermodynamic efficiency of the process or cycle.

Three basic differences between PEC for single-phase heat exchangers and PEC for two-phase heat exchangers are as follows:

(a) LMTD is affected by two-phase pressure drop.
(b) Work produced or consumed by turbine or compressor is the key point of concern; the work consumed by pump or fan is negligible.
(c) LMTD influences the heat exchanger size or the thermodynamic efficiency of the heat-actuated system and the effect of the two-phase fluid pressure drop on LMTD is the point of concern.

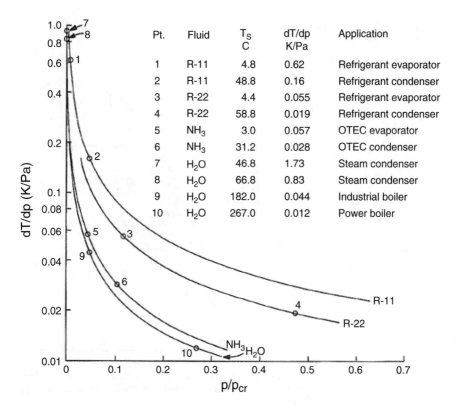

Pt.	Fluid	T_S C	dT/dp K/Pa	Application
1	R-11	4.8	0.62	Refrigerant evaporator
2	R-11	48.8	0.16	Refrigerant condenser
3	R-22	4.4	0.055	Refrigerant evaporator
4	R-22	58.8	0.019	Refrigerant condenser
5	NH_3	3.0	0.057	OTEC evaporator
6	NH_3	31.2	0.028	OTEC condenser
7	H_2O	46.8	1.73	Steam condenser
8	H_2O	66.8	0.83	Steam condenser
9	H_2O	182.0	0.044	Industrial boiler
10	H_2O	267.0	0.012	Power boiler

Fig. 4.2 dT/dp vs. p/p_0 [1, 2]

Table 4.1 shows PEC corresponding to the FG, FN and VG cases for the two-phase heat exchanger applications. This table is equally applicable for vaporisation or condensation inside tubes or on the outside of tubes in a bundle.

Heat transfer coefficient and friction factor (or Δp characteristics) of the enhanced tube as a function of the flow and geometry conditions must be known before PEC calculations are done. The heat transfer and friction characteristics for vaporisation or condensation on enhanced surfaces depend on the following:

(a) Enhanced surface geometry
(b) Operating or saturation pressure
(c) The mass velocity (G), vapour quality (x), heat flux (q) or wall-to-fluid temperature difference ($T_w - T_s$)

The data may be measured as a function of the local vapour quality or as average values over a given inlet and exit vapour quality. Jensen [3] gives some numerical examples of two-phase heat transfer enhancement PEC calculations.

Webb [4] studied the PEC for single-phase flow as well as two-phase heat exchanger at a broad level. He investigated the PEC for performance improvement

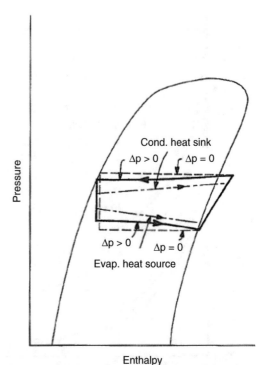

Enthalpy

Fig. 4.3 Pressure vs. enthalpy diagram for a refrigeration cycle showing the effect of pressure drop on the LMTD in each heat exchanger [1, 2]

Table 4.1 Performance evaluation criteria for two-phase heat exchange system

Case	Geometry	Fixed parameters				Objective
		W	P_{w}	Q	ΔT_i^n	
FG-1a	N,L			X		$\uparrow Q$
FG-1b	N,L		X	X	X	$\downarrow \Delta T_i*$
FG-3	N,L	X		X	X	$\downarrow P_{\mathrm{w}}^b$
FN-1	N	X		X	X	$\downarrow L$
FN-2	N	X		X	X	$\downarrow L$
FN-3	N	X		X		$\downarrow P_{\mathrm{w}}^b$
VG-1		X	X	X		$\downarrow NL$
VG-2a	NL		X		X	$\uparrow Q$
VG-2b	NL		X		X	$\downarrow \Delta T_i$
VG-3	NL	X		X		$\downarrow P_{\mathrm{w}}^b$

of heat exchanger and considered four design cases: (1) reduced heat exchanger material, (2) increased heat duty, (3) reduced log mean temperature difference and (4) reduced pumping power. He also evaluated the performance evaluation criteria for fixed flow area and variable flow area.

Table 4.2 Thermo-economically optimised parameters of condenser and evaporator [5]

	Condenser	Evaporator
Input		
Refrigerant	R-22	R-22
Heat load, kW	21.1	17.6
Single-phase fluid medium	Water	Air
Its exit/inlet temperature, °C	35.0/29.4	17.2/26.7
α, exit/inlet absolute temperature ratio	1.0185	0.9683
$\dot{C}_{min}$, minimum heat capacity, kW/°C	3.768	1.853
Reference temperature, K	294.4	294.4
Operating hours, h/year	5000	5000
Tube diameters, o.d./i.d., mm	15.88/14.10	15.88/14.10
No. of tube passes/unit cells	6	5*
λ_{H}, unit cost of heat transfer, $/kW-h	0.08	0.10
λ_{A}, unit cost of moving single-phase fluid, $/kW-h	0.04	0.04
$\dot{z}_{m}$, modified annualised cost, $/m²-year	13.40	18.84
$\dot{z}_{m}$, expressed as hourly value based on 5000 h/year operation, $/m²-h	0.00268	0.00377
Output		
V^{*}, optimum velocity, m/h	3660	4100
U^{*}, overall heat transfer coefficient, kW/m²-C	0.9576	1.216
γ_{UA}, cost per unit overall conductance, $-°C/kW-h	2.93×10^{-3}	3.51×10^{-3}
β, dimensionless ratio		
$\beta \underline{\Delta} (\lambda_{H} T_{0}/\gamma_{UA})$	8038	8382
$\beta_{C} \underline{\Delta} (\beta/\alpha)(\alpha - 1)^{2}$	2.701	8.699
N_{tu}^{*}, optimum number of transfer units	1.48	2.39
A^{*}, optimum surface area, m²	5.82	3.64
ϵ, temperature effectiveness	0.7724	0.908
T_{c} or T_{e}, temperature of refrigerant, °C	36.7	16.2
Approximate V^{*}, m/h	7876	6221

Additional input parameters specified for the unit cell of finned air-cooled evaporator
Number of fins = 5.9 fins/cm, fin thickness = 0.15 mm, in-line arrangement
Number of tube rows in direction of airflow = 3
Number of tube rows in direction of refrigerant flow = 4
Transverse/longitudinal pitch = 44.5/50.8 mm

Two-phase heat exchangers such as condenser or evaporator have been thermo-economically optimised based on second law heat generation analysis by Zubair et al. [5]. Thermodynamically optimised input and output variables for improving performance of the heat exchanger are presented in Table 4.2. They gave a term "internal economy" which was meant to estimate the economic value of entropy generated in the heat exchanger again its capital investment. They experimentally found the results in terms of optimum heat exchanger area as a function of temperatures of coolant, unit cost of energy dissipated and overall heat transfer coefficient. They observed that cost and power requirement for circulation of single-phase fluid

Table 4.3 Equations used in estimation of irreversibilities (lost work) associated with pressure drop of single-phase fluid [5]

	$(\Delta p)_{sp}/\rho$		Friction factor	$T_0\dot{S}_A$
Liquid heat exchanger (internal flow)	$\frac{fLV^2}{2g_cD}$		$\frac{f}{8}=a\mathrm{Re}^{-b}$	K'_vAV^{3-b}
Gas/air heat exchanger (external flow)	$\frac{G^2_{max}}{2g_c\rho^2}\frac{L}{D_H}\left(\frac{D_H}{S_T}\right)^{k1}\left(\frac{S_L}{S_T}\right)^{k2}f'$		$f'=c\,\mathrm{Re}^{-d}$	$K''_vAV^{3-d}_{max}$
$\mathrm{Re}=\frac{\rho VD}{\mu}$	For turbulent internal flow	$a=0.023$ $m=3-b$	$b=0.2$ after Colburn $n=1-b$	
$\mathrm{Re}=\frac{G_{max}D_H}{\mu}$	For in-line tube arrangement external flow	$c=1.92$ $k1=0.4$ $m=3-d$	$d=0.115$ $k2=0.6$ $n=1-d$	

$D_H=4\frac{\text{Free volume in tube bank}}{\text{Exposed surface area of tube}}\;K_v=\lambda_AK'_v\,\text{or}\,\lambda_AK''_v$

$K'_v=2a\frac{T_0}{T_{A,\,\text{In}}}\frac{\rho}{2g_c}\left(\frac{\rho D}{\mu}\right)^{-b};K''_v=c\frac{T_0}{T_{A,\,\text{In}}}\frac{\rho}{2g_c}\frac{A_c}{A}\left(\frac{L}{D_H}\right)\left(\frac{D_H}{S_T}\right)^{k1}\left(\frac{S_L}{S_T}\right)^{k2}\left(\frac{\rho D_H}{\mu}\right)^{-d}$

was expensive. Equations used in the estimation of irreversibility (lost work), pressure drop and friction factor associated with the single-phase are presented in Table 4.3. London and Shah [6] proposed the two-phase pressure drop model for evaluating the irreversibility.

More information on performance evaluation criteria for two-phase heat transfer augmentation techniques can be obtained from Webb [1, 2, 7], Reay [8], Royal and Bergles [9], Lord et al. [10] and Withers and Habdas [11], Webb [12].

References

1. Webb RL (1988) Performance evaluation criteria for enhanced surface geometries used in two-phase heat exchangers. In: Shah RK, Subbarao EC, Mashelkar RA (eds) Heat transfer equipment design. Hemisphere, Washington, DC, pp 697–706
2. Webb RL (1988) Performance evaluation criteria for enhanced tube geometries used in two-phase heat exchangers. In: Shah RK, Subbarao EC, Mashelkar RA (eds) Heat transfer equipment design. Hemisphere, Washington, DC, pp 697–704
3. Jensen MK (1988) Enhanced forced convective vaporization and condensation inside tubes. In: Shah RK, Subbarao EC, Mashelkar RA (eds) Heat transfer equipment design. Hemisphere, Washington, DC, pp 681–696
4. Webb RL (1981) Performance evaluation criteria for use of enhanced heat transfer surfaces in heat exchanger design. Int J Heat Mass Transfer 24(4):715–726
5. Zubair SM, Kadaba PV, Evans RB (1987) Second-law-based thermoeconomic optimization of two-phase heat exchangers. J Heat Transfer 109(2):287–294
6. London AL, Shah RK (1983) Costs of irreversibilities in heat exchanger design. Heat Transfer Eng 4(2):59–73
7. Webb RL (1994) Principles of enhanced heat transfer. John Wiley, New York

8. Reay DA (1991) Heat transfer enhancement—a review of techniques and their possible impact on energy efficiency in the U.K. Heat Recov Syst CHP 11(1):40

9. Royal JH, Bergles AE (1978) Augmentation of horizontal in-tube condensation by means of twisted tape inserts and internally finned tubes. J Heat Transfer 100:17–24

10. Lord RG, Bussjager RC, Geary DF (1978) High performance heat exchanger, U.S. Patent 4 (118):944

11. Withers JG, Habdas EP (1974) Heat transfer characteristics of helical corrugated tubes for in-tube boiling of refrigerants. AIChe Symp Ser 70(138):98–106

12. Webb RL, Gupte NS (1992) A critical review of correlations for convective vaporization in tubes and tube banks. Heat Transfer Eng 13(3):58–81

Chapter 5
Conclusions

- This Springer Brief has started with the discussion on the importance of performance evaluation criteria in the design of heat exchangers for single-phase flow and two-phase flow. The objectives and constraints in developing different performance evaluation criteria have been discussed.
- Then attention has been mainly focused on performance evaluation criteria for single-phase flow. The objectives and constraints of fixed geometry criteria, variable geometry criteria and fixed cross-sectional flow area criteria have been discussed. The thermal resistance, St and f relations, heat exchanger effectiveness, effect of reduced exchanger flow rate and flow over finned tube banks have been presented.
- Thereafter discussion follows with works carried out by various researchers across the globe on performance evaluation criteria based on first law analysis and second law analysis. The importance of using proper criteria which considers both first law and second law analysis in order to evaluate the performance of augmentation techniques has been presented.
- The Brief concludes with performance evaluation criteria analysis for two-phase flow. The disadvantages of using single-phase PEC correlations for two-phase flow have been discussed. Also, the factors to be considered to modify the single-phase PEC such that they can be applicable to two-phase flow have been listed.

© The Author(s), under exclusive license to Springer Nature Switzerland AG 2020   107
S. K. Saha et al., *Performance Evaluation Criteria in Heat Transfer Enhancement*,
SpringerBriefs in Applied Sciences and Technology,
https://doi.org/10.1007/978-3-030-20758-8_5

Index

|||| || || ||||||||||| ||||| ||||| ||| ||||| ||||| || |||